THE NATIONAL ACADEMIES

National Academy of Sciences
National Academy of Engineering
Institute of Medicine
National Research Council

The **National Academy of Sciences** is a private, nonprofit, self-perpetuating society of distinguished scholars engaged in scientific and engineering research, dedicated to the furtherance of science and technology and to their use for the general welfare. Upon the authority of the charter granted to it by the Congress in 1863, the Academy has a mandate that requires it to advise the federal government on scientific and technical matters. Dr. Bruce M. Alberts is president of the National Academy of Sciences.

The **National Academy of Engineering** was established in 1964, under the charter of the National Academy of Sciences, as a parallel organization of outstanding engineers. It is autonomous in its administration and in the selection of its members, sharing with the National Academy of Sciences the responsibility for advising the federal government. The National Academy of Engineering also sponsors engineering programs aimed at meeting national needs, encourages education and research, and recognizes the superior achievements of engineers. Dr. William A. Wulf is president of the National Academy of Engineering.

The **Institute of Medicine** was established in 1970 by the National Academy of Sciences to secure the services of eminent members of appropriate professions in the examination of policy matters pertaining to the health of the public. The Institute acts under the responsibility given to the National Academy of Sciences by its congressional charter to be an adviser to the federal government and, upon its own initiative, to identify issues of medical care, research, and education. Dr. Kenneth I. Shine is president of the Institute of Medicine.

The **National Research Council** was organized by the National Academy of Sciences in 1916 to associate the broad community of science and technology with the Academy's purposes of furthering knowledge and advising the federal government. Functioning in accordance with general policies determined by the Academy, the Council has become the principal operating agency of both the National Academy of Sciences and the National Academy of Engineering in providing services to the government, the public, and the scientific and engineering communities. The Council is administered jointly by both Academies and the Institute of Medicine. Dr. Bruce M. Alberts and Dr. William A. Wulf are chairman and vice chairman, respectively, of the National Research Council.

METHYL BROMIDE RISK CHARACTERIZATION IN CALIFORNIA

Subcommittee for the Review of the Risk Assessment
of Methyl Bromide

Committee on Toxicology

Board on Environmental Studies and Toxicology

Commission on Life Sciences

National Research Council

NATIONAL ACADEMY PRESS
Washington, D.C.

NATIONAL ACADEMY PRESS   2101 Constitution Ave., N.W.   Washington, D.C. 20418

NOTICE: The project that is the subject of this report was approved by the Governing Board of the National Research Council, whose members are drawn from the councils of the National Academy of Sciences, the National Academy of Engineering, and the Institute of Medicine. The members of the committee responsible for the report were chosen for their special competences and with regard for appropriate balance.

This project was supported by Agreement No. 98-0316 between the National Academy of Sciences and the California Department of Pesticide Regulation. Any opinions, findings, conclusions, or recommendations expressed in this publication are those of the author(s) and do not necessarily reflect the view of the organizations or agencies that provided support for this project.

International Standard Book Number 0-309-07087-2

Additional copies of this report are available from:

National Academy Press
2101 Constitution Ave., NW
Box 285
Washington, DC 20055

800-624-6242
202-334-3313 (in the Washington metropolitan area)
http://www.nap.edu

Copyright 2000 by the National Academy of Sciences. All rights reserved.

Printed in the United States of America

SUBCOMMITTEE ON METHYL BROMIDE

CHARLES H. HOBBS *(Chair)*, Lovelace Respiratory Research Institute, Albuquerque, New Mexico
JANICE E. CHAMBERS, Mississippi State University, College of Veterinary Medicine, Mississippi State, Mississippi
FRANK N. DOST, Professor Emeritus, Oregon State University, Corvallis, Oregon
DALE B. HATTIS, Clark University, Worcester, Massachusetts
MATTHEW C. KEIFER, University of Washington, Seattle, Washington
ULRIKE LUDERER, University of California-Irvine, Irvine, California
GLENN C. MILLER, University of Nevada, Reno, Nevada
SYLVIA S. TALMAGE, Oak Ridge National Laboratory, Oak Ridge, Tennessee

Staff

ROBERTA M. WEDGE, Project Director
EILEEN N. ABT, Research Associate
ROBERT J. CROSSGROVE, Editor
MIRSADA KARALIC-LONCAREVIC, Information Specialist
LUCY V. FUSCO, Project Assistant

Sponsor

CALIFORNIA DEPARTMENT OF PESTICIDE REGULATION

COMMITTEE ON TOXICOLOGY

BAILUS WALKER, JR. *(Chair)*, Howard University Medical Center, Washington, D.C.
MELVIN E. ANDERSEN, Colorado State University, Fort Collins, Colorado
GERMAINE M. BUCK, State University of New York at Buffalo
GARY P. CARLSON, Purdue University, West Lafayette, Indiana
JACK H. DEAN, Sanofi Pharmaceuticals, Inc., Malverne, Pennsylvania
ROBERT E. FORSTER II, University of Pennsylvania, Philadelphia, Pennsylvania
PAUL M.D. FOSTER, Chemical Industry Institute of Toxicology, Research Triangle Park, North Carolina
DAVID W. GAYLOR, U.S. Food and Drug Administration, Jefferson, Arkansas
JUDITH A. GRAHAM, U.S. Environmental Protection Agency, Research Triangle Park, North Carolina
SIDNEY GREEN, Howard University, Washington, D.C.
WILLIAM E. HALPERIN, National Institute for Occupational Safety and Health, Cincinnati, Ohio
CHARLES H. HOBBS, Lovelace Respiratory Research Institute and Lovelace Biomedical and Environmental Research Institute, Albuquerque, New Mexico
FLORENCE K. KINOSHITA, Hercules Incorporated, Wilmington, Delaware
MICHAEL J. KOSNETT, University of Colorado Health Sciences Center, Denver, Colorado
MORTON LIPPMANN, New York University School of Medicine, Tuxedo, New York
THOMAS E. MCKONE, University of California, Berkeley, California
ERNEST E. MCCONNELL, ToxPath, Inc., Raleigh, North Carolina
DAVID H. MOORE, Battelle Memorial Institute, Bel Air, Maryland
GÜNTER OBERDÖRSTER, University of Rochester, Rochester, New York
JOHN L. O'DONOGHUE, Eastman Kodak Company, Rochester, New York
GEORGE M. RUSCH, AlliedSignal, Inc., Morristown, New Jersey
MARY E. VORE, University of Kentucky, Lexington, Kentucky
ANNETTA P. WATSON, Oak Ridge National Laboratory, Oak Ridge, Tennessee

Staff

KULBIR S. BAKSHI, Program Director
SUSAN N.J. PANG, Program Officer
ABIGAIL STACK, Program Officer
RUTH E. CROSSGROVE, Publications Manager
KATHRINE J. IVERSON, Manager, Toxicology Information Center
LUCY V. FUSCO, Project Assistant
LEAH PROBST, Project Assistant

BOARD ON ENVIRONMENTAL STUDIES AND TOXICOLOGY

GORDON ORIANS (*Chair*), University of Washington, Seattle, Washington
DONALD MATTISON (*Vice Chair*), March of Dimes, White Plains, New York
DAVID ALLEN, University of Texas, Austin, Texas
INGRID C. BURKE, Colorado State University, Fort Collins, Colorado
WILLIAM L. CHAMEIDES, Georgia Institute of Technology, Atlanta, Georgia
JOHN DOULL, University of Kansas Medical Center, Kansas City, Kansas
CHRISTOPHER B. FIELD, Carnegie Institute of Washington, Stanford, California
JOHN GERHART, University of California, Berkeley, California
J. PAUL GILMAN, Celera Genomics, Rockville, Maryland
BRUCE D. HAMMOCK, University of California, Davis, California
MARK HARWELL, University of Miami, Miami, Florida
ROGENE HENDERSON, Lovelace Respiratory Research Institute, Albuquerque, New Mexico
CAROL HENRY, Chemical Manufacturers Association, Arlington, Virginia
BARBARA HULKA, University of North Carolina, Chapel Hill, North Carolina
JAMES F. KITCHELL, University of Wisconsin, Madison, Wisconsin
DANIEL KREWSKI, University of Ottawa, Ottawa, Ontario
JAMES A. MACMAHON, Utah State University, Logan, Utah
MARIO J. MOLINA, Massachusetts Institute of Technology, Cambridge, Massachusetts
CHARLES O'MELIA, Johns Hopkins University, Baltimore, Maryland
WILLEM F. PASSCHIER, Health Council of the Netherlands
KIRK SMITH, University of California, Berkeley, California
MARGARET STRAND, Oppenheimer Wolff Donnelly & Bayh, LLP, Washington, D.C.
TERRY F. YOSIE, Chemical Manufacturers Association, Arlington, Virginia

Senior Staff

JAMES J. REISA, Director
DAVID J. POLICANSKY, Associate Director and Senior Program Director for Applied Ecology
CAROL A. MACZKA, Senior Program Director for Toxicology and Risk Assessment
RAYMOND A. WASSEL, Senior Program Director for Environmental Sciences and Engineering
KULBIR BAKSHI, Program Director for the Committee on Toxicology
LEE R. PAULSON, Program Director for Resource Management
ROBERTA M. WEDGE, Program Director for Risk Analysis

COMMISSION ON LIFE SCIENCES

MICHAEL T. CLEGG *(Chair)*, University of California, Riverside, California
PAUL BERG *(Vice Chair)*, Stanford University, Stanford, California
FREDERICK R. ANDERSON, Cadwalader, Wickersham & Taft, Washington, D.C.
JOANNA BURGER, Rutgers University, Piscataway, New Jersey
JAMES E. CLEAVER, University of California, San Francisco, California
DAVID S. EISENBERG, University of California, Los Angeles, California
JOHN L. EMMERSON, Fishers, Indiana
NEAL L. FIRST, University of Wisconsin, Madison, Wisconsin
DAVID J. GALAS, Keck Graduate Institute of Applied Life Science, Claremont, California
DAVID V. GOEDDEL, Tularik, Inc., South San Francisco, California
ARTURO GOMEZ-POMPA, University of California, Riverside, California
COREY S. GOODMAN, University of California, Berkeley, California
JON W. GORDON, Mount Sinai School of Medicine, New York, New York
DAVID G. HOEL, Medical University of South Carolina, Charleston, South Carolina
BARBARA S. HULKA, University of North Carolina, Chapel Hill, North Carolina
CYNTHIA J. KENYON, University of California, San Francisco, California
BRUCE R. LEVIN, Emory University, Atlanta, Georgia
DAVID M. LIVINGSTON, Dana-Farber Cancer Institute, Boston, Massachusetts
DONALD R. MATTISON, March of Dimes, White Plains, New York
ELLIOT M. MEYEROWITZ, California Institute of Technology, Pasadena, California
ROBERT T. PAINE, University of Washington, Seattle, Washington
RONALD R. SEDEROFF, North Carolina State University, Raleigh, North Carolina
ROBERT R. SOKAL, State University of New York, Stony Brook, New York
CHARLES F. STEVENS, The Salk Institute for Biological Studies, La Jolla, California
SHIRLEY M. TILGHMAN, Princeton University, Princeton, New Jersey
RAYMOND L. WHITE, University of Utah, Salt Lake City, Utah

Staff

WARREN R. MUIR, Executive Director
JACQUELINE K. PRINCE, Financial Officer
BARBARA B. SMITH, Administrative Associate
LAURA T. HOLLIDAY, Senior Program Assistant

OTHER REPORTS OF THE
BOARD ON ENVIRONMENTAL STUDIES AND TOXICOLOGY

Copper in Drinking Water (2000)
Ecological Indicators for the Nation (2000)
Waste Incineration and Public Health (1999)
Hormonally Active Agents in the Environment (1999)
Research Priorities for Airborne Particulate Matter: I. Immediate Priorities and a Long-Range Research Portfolio (1998); II. Evaluating Research Progress and Updating the Portfolio (1999)
Ozone-Forming Potential of Reformulated Gasoline (1999)
Risk-Based Waste Classification in California (1999)
Arsenic in Drinking Water (1999)
Brucellosis in the Greater Yellowstone Area (1998)
The National Research Council's Committee on Toxicology: The First 50 Years (1997)
Toxicologic Assessment of the Army's Zinc Cadmium Sulfide Dispersion Tests (1997)
Carcinogens and Anticarcinogens in the Human Diet (1996)
Upstream: Salmon and Society in the Pacific Northwest (1996)
Science and the Endangered Species Act (1995)
Wetlands: Characteristics and Boundaries (1995)
Biologic Markers (5 reports, 1989-1995)
Review of EPA's Environmental Monitoring and Assessment Program (3 reports, 1994-1995)
Science and Judgment in Risk Assessment (1994)
Ranking Hazardous Waste Sites for Remedial Action (1994)
Pesticides in the Diets of Infants and Children (1993)
Issues in Risk Assessment (1993)
Setting Priorities for Land Conservation (1993)
Protecting Visibility in National Parks and Wilderness Areas (1993)
Dolphins and the Tuna Industry (1992)
Hazardous Materials on the Public Lands (1992)
Science and the National Parks (1992)
Animals as Sentinels of Environmental Health Hazards (1991)
Assessment of the U.S. Outer Continental Shelf Environmental Studies Program, Volumes I-IV (1991-1993)
Human Exposure Assessment for Airborne Pollutants (1991)
Monitoring Human Tissues for Toxic Substances (1991)
Rethinking the Ozone Problem in Urban and Regional Air Pollution (1991)
Decline of the Sea Turtles (1990)

Copies of these reports may be ordered from the National Academy Press
(800) 624-6242 or (202) 334-3313
www.nap.edu

Preface

One of the most widely used pesticides in California is methyl bromide, a gaseous fumigant that is used on a variety of crops primarily as a preplant soil insecticide, on post-harvest commodities, and in some residences as a fumigant. Although methyl bromide is a recognized stratospheric ozone depleter and is scheduled to be phased out completely by 2005 under the United Nations Montreal Protocol, it continues to be of concern for the health of agricultural workers and exposed residents.

The California Department of Pesticide Regulation (DPR) is responsible for the development of regulations that determine the site-specific permit conditions for the application of pesticides in the state. California is currently in the process of proposing new regulations for issuing methyl bromide permits that require submission of a worksite plan by the property operator, provide extra protection for children in nearby schools, establish minimum buffer zones around application sites, require that nearby residents receive prior notification of the application of methyl bromide, and set new limits on hours that fumigation employees may work. To develop these regulations, the DPR prepared a risk-characterization document to evaluate the toxicity and exposure potential for workers and residents resulting from the inhalation of this pesticide.

Under Section 57004 of the California Health and Safety Code, the scientific basis of the proposed regulations is subject to external peer review by the National Academy of Sciences, the University of California, or other similar institution of higher learning or group of scientists. This report addresses that regulatory requirement by reviewing the DPR risk-characterization document that supports the proposed regulations.

The National Research Council (NRC), the operating arm of The National Academies, assigned the task of preparing this report to its Committee on Toxicology, which convened the subcommittee for the review of the risk assessment of methyl bromide. The subcommittee was charged with the following tasks: (1) determine whether all relevant data were considered, (2) determine the appropriateness of the critical studies and endpoints used in the risk assessment and in the derivation of exposure limits, (3) consider the mode of action of methyl bromide and its implications in risk assessment, and (4) determine the appropriateness of the exposure assessment and mathematical models used. The subcommittee also identified data gaps and made recommendations for further research relevant to setting exposure limits for methyl bromide.

To prepare this report, the subcommittee reviewed the materials supplied by DPR, additional supporting materials received from other individuals and organizations, and the information gathered at a public meeting held in Irvine, California, on October 4, 1999. The subcommittee wishes to thank the following members of the California Department of Pesticide Regulation—Paul Gosselin, Acting Chief Deputy Director, Lori Lim, and Thomas Thongsinthusak—for providing the subcommittee with information on methyl bromide toxicology and exposure data and models, for their presentation at the public meeting, and for responding to follow up requests from the subcommittee members. We also gratefully acknowledge Vincent J. Piccirillo, NPC, Inc., Bill Walker, Environmental Working Group, and Amy Kyle, Consulting Scientist, for providing background information and for making presentations to the subcommittee, and Jodi Kuhn, Methyl Bromide Industry Panel of the Chemical Manufacturers Association, for providing background materials as well.

This report has been reviewed in draft form by individuals chosen for their diverse perspectives and technical expertise, in accordance with procedures for reviewing NRC reports approved by the NRC's Report Review Committee. The purpose of this independent review is to provide candid and critical comments that will assist NRC in making the published report as sound as possible and to ensure that the report meets institutional standards for objectivity, evidence, and responsiveness to the study charge. The content of the final report is the responsibility of NRC and the study subcommittee, and not the responsibility of the reviewers. The review comments and draft manuscript remain confidential to protect the integrity of the deliberative process. We wish to thank the following individuals, who are neither officials nor employees of NRC, for their participation in the review of the report: Dana Barr, Centers for Disease Control and Prevention; David Dorman, Chemical Industry Institute of Toxicology; David Gaylor, National Center for Toxicological Research; Craig Harris, University of Michigan; John Morris, University of Connecticut; and P. Barry Ryan, Emory University. These reviewers have provided many constructive comments and suggestions; it must be emphasized, however, that responsibility for the final content of this report rests entirely with the authoring subcommittee and NRC.

I am also grateful for the assistance of NRC staff in the preparation of this report. In particular, the subcommittee wishes to acknowledge Roberta Wedge, staff officer for the subcommittee, and Eileen Abt, research associate, with the Board on Environmental Studies and Toxicology. Other staff members who contributed to this effort are Robert Crossgrove, editor, Lucy Fusco, project assistant, and Kulbir Bakshi, program director for the Committee on Toxicology.

Finally, I would like to thank the members of the subcommittee for their valuable expertise and dedicated efforts throughout the preparation of this report. Their efforts are much appreciated.

Charles H. Hobbs, D.V.M.
Chair, Subcommittee on the Review of
the Risk Assessment for Methyl Bromide

Contents

EXECUTIVE SUMMARY 1

1 INTRODUCTION 8
   Background, 8
   California Regulations, 9
   The Subcommittee's Task, 10
   Organization of the Report, 11

2 TOXICOLOGY AND HAZARD IDENTIFICATION 12
   Pharmacokinetics, 13
   Genotoxicity, 16
   Acute Toxicity, 17
   Subchronic Toxicity, 18
   Chronic Inhalation and Oncogenicity, 20
   Reproductive Toxicity, 23
   Developmental Toxicity, 27
   Neurotoxicity, 31
   Selection of Critical Effects for Acute Toxicity, 32

3 EXPOSURE ASSESSMENT 35
   Background, 35
   Likely Exposure Scenarios, 36
   Quality of Data Available for Characterizing Exposures, 39
   Accuracy and Appropriateness of Available Modeling Tools, 46
   Summary, 54

4 RISK CHARACTERIZATION 56
   Risk Characterization Goals, 56
   Hazard Identification, 57
   Exposure Assessment, 60
   Margin-of-Exposure Analysis, 64
   Uncertainty Issues, 66
   Summary, 69

5 CONCLUSIONS AND RECOMMENDATIONS 70
   Toxicological Information, 70
   Exposure Assessment, 72
   Risk Characterization, 74

REFERENCES 76

APPENDIX A
Biographical Information on the Subcommittee for the
Review of the Risk Assessment of Methyl Bromide 84

APPENDIX B
Public Access Materials 87

APPENDIX C
Calculation of Air Exchange Rates 90

Methyl Bromide Risk Characterization in California

Executive Summary

Methyl bromide is a gaseous pesticide used to fumigate soil, crops, commodity warehouses, and commodity-shipping facilities. Up to 17 million pounds of methyl bromide are used annually in California to treat grapes, almonds, strawberries, and other crops. Methyl bromide is also a known stratospheric ozone depleter and, as such, is scheduled to be phased out of use in the United States by 2005 under the United Nations Montreal Protocol.

In California, the use of methyl bromide is regulated by the Department of Pesticide Regulation (DPR), which is responsible for establishing the permit conditions that govern the application of methyl bromide for pest control. The actual permits for use are issued on a site-specific basis by the local county agricultural commissioners. Because of concern for potential adverse health effects, in 1999 DPR developed a draft risk characterization document for inhalation exposure to methyl bromide. The DPR document is intended to support new regulations regarding the agricultural use of this pesticide. The proposed regulations encompass changes to protect children in nearby schools, establish minimum buffer zones around application sites, require notification of nearby residents, and set new limits on hours that fumigation employees may work.

THE SUBCOMMITTEE'S TASK

The State of California requires that DPR arrange for an external peer review of the scientific basis for all regulations. To this end, the National Research Council (NRC) was asked to review independently the draft risk characterization document prepared by DPR for inhalation exposure to methyl bromide. NRC assigned the task to the Committee on Toxicology, which convened the Subcommittee for the Review of the Risk Assessment of Methyl Bromide. The subcommittee was asked to review the data, determine the appropriateness of the critical studies, consider the mode of action of methyl bromide and its implications in risk assessment, determine the appropriateness of the exposure assessment and the mathematical models, and identify data gaps and make recommendations for further research.

THE SUBCOMMITTEE'S EVALUATION

The 1999 risk characterization document prepared by DPR is a revision of a 1992 preliminary risk assessment that addressed acute inhalation exposure of residents reentering fumigated homes. The 1999 document updates the toxicity information on methyl bromide and provides a more extensive review of the worker and residential exposure data gathered by the methyl bromide manufacturers and applicators and DPR itself over the past several years. The toxicity and exposure data were combined to establish margins of exposure[1] for agricultural workers, residents living near fumigated fields, and residents reentering fumigated homes. The subcommittee's comments on the DPR risk characterization document and its recommendations for further studies are summarized below under three broad categories: toxicology, exposure assessment, and risk characterization.

Toxicology

The DPR risk characterization document presents information on the toxicokinetics and toxicity of methyl bromide, including its acute, subchronic, chronic, developmental, reproductive, neurological, and genotoxic effects. The subcommittee agrees with DPR that the critical target organ for acute exposure to methyl bromide is the nervous system. Methyl bromide also appears to be a developmental and possibly a reproductive toxicant.

The DPR report appropriately summarizes the available toxicokinetic data on methyl bromide in terms of its absorption, distribution and excretion, but it provides only a limited discussion of the metabolism of the pesticide. That discussion is particularly important, because in some individuals there appears

---

[1] Margin of exposure is a ratio of the concentration at which adverse effects occur to the estimate of concentration found in the workplace or ambient air.

to be a more effective reaction between methyl bromide and glutathione transferase, which can alter the sensitivity of those individuals to its toxic effects. Although DPR adequately reviewed the available literature on the genotoxicity of methyl bromide, it failed to elucidate the relationship between the mutagenicity of methyl bromide and its potential carcinogenicity.

When possible, the DPR report identified the no-observed-adverse-effect level (NOAEL)[2] or the lowest-observed-adverse-effect level (LOAEL)[3] following acute, subchronic, and chronic exposures. The subcommittee agrees with the critical studies and NOAELs selected by DPR in developing the reference concentrations for acute, subchronic, and chronic exposures. For acute toxicity, DPR chose 40 parts per million (ppm) as a NOAEL based on a developmental toxicity study in which rabbits exposed in utero exhibited fused breastbones and gallbladder agenesis (lack of gallbladder development). This NOAEL resulted in an acute inhalation reference concentration[4] (RfC) of 210 parts per billion (ppb) for humans. For subchronic toxicity, 1-week and 6-week RfCs were derived. The subcommittee supports DPR's 1-week RfC of 120 ppb and 70 ppb for adults and children, respectively, based on a NOAEL of 20 ppm for convulsions, paresis, and death in pregnant rabbits. The subcommittee also supports DPR's 6-week RfCs of 2 ppb and 1 ppb for adults and children, respectively, based on a LOAEL of 5 ppm for decreased responsiveness and spleen weight in dogs. The subcommittee notes that the reported neurotoxic effects (lack of responsiveness in two of eight dogs) seen in the subchronic toxicity dog study used to derive the 6-week RfC, because the observations were not part of the study protocol and were not dramatic. Nevertheless, the subcommittee believes that the reported effects may be indicative of neurotoxicity.

DPR used a 29-month rat inhalation study for the derivation of the chronic exposure RfC of 2 ppb for adults and 1 ppb for children. The LOAEL of 3 ppm identified in that study was based on an increase in the number of cells and a change in cell type and function in the nasal cavity. The subcommittee

---

[2]NOAEL is an exposure level at which there are no statistically or biologically significant increases in the frequency or severity of adverse effects between the exposed population and its appropriate control.

[3]LOAEL is the lowest exposure level at which there are statistically or biologically significant increases in frequency or severity of adverse effects between the exposed population and its appropriate control.

[4]A reference concentration is an estimate of the concentration of a substance that is unlikely to cause noncancer health effects in humans during a lifetime. It is used by DPR as a regulatory value for establishing buffer zones to protect residents from adverse effects of methyl bromide exposure.

notes that although the effects seen in the adult rats were dose-related and statistically significant, they were slight or equivocal and only observed in aged rats. Nevertheless, the subcommittee agrees with DPR that this is the correct study to use for a chronic exposure RfC.

DPR concluded that 3 ppm was a NOAEL in two rat reproductive toxicity studies, although the subcommittee questions whether the reduction in fertility observed in the F1 generation was of reproductive origin or developmental origin. Studies conducted in rats and rabbits indicate that in utero exposure to methyl bromide results in developmental toxicity. In rats this was manifested by reduced body and brain weights in pups and reduced fertility in gestationally exposed offspring. In rabbit offspring, gallbladder agenesis, reduced fetal weights, and increased frequency of fused sternebrae were seen. Although the subcommittee recognizes that any of those effects individually might be considered equivocal, together they suggest that methyl bromide has the potential to be a developmental toxicant.

Recommendations

- Studies should be conducted to confirm the neurotoxic effects seen in dogs following subchronic exposures.
- The developmental and reproductive effects of methyl bromide should be further investigated to determine whether it is a direct-acting reproductive toxicant or a developmental toxicant to the reproductive system, whether methyl bromide is excreted in breast milk, and whether the gallbladder agenesis seen in offspring occurs following a single exposure during a critical period of development.
- Neurological testing of workers should be conducted to determine possible long-term or permanent effects following occupational exposure to methyl bromide.
- DPR should review the literature on methyl bromide and other methylating agents to assist them in understanding why methyl bromide, an in vitro mutagen, is not an in vivo carcinogen. This might also help elucidate the mechanism of methyl bromide toxicity.

Exposure

The DPR report presents a substantial amount of data on exposure estimates for a wide variety of worker and residential exposure scenarios. The majority of the exposure information is obtained from studies that were conducted to establish permit conditions. This information was not always collected in a consistent and comprehensive manner, or in compliance with Good Laboratory Practices.

The DPR report addresses (1) exposures of workers, (2) exposures of individuals due to environmental transport of methyl bromide away from the site of direct application, and (3) exposures of residents returning to fumigated houses. DPR focuses principally on occupational exposure scenarios, presenting data on 160 exposure categories, with estimates for acute (daily), short-term (7 days), seasonal (90 days), and chronic (annual) exposures. The occupational exposures range widely from a high of 8,458 ppb to a low of 0.6 ppb. DPR provides data from two studies on exposures of residents of houses neighboring fumigated structures. No air sampling was conducted to assess individual exposures near commodity fumigation sites; DPR assumes that people are exposed to concentrations of 210 ppb. Exposure data on residents returning to fumigated homes are taken from five houses in southern California. Exposure modeling and field studies indicate that some worker exposures exceed protective levels by more than an order of magnitude, whereas potential exposures of residents living near fumigated fields and facilities are unquantified.

Although DPR compiled a large quantity of exposure data in its document, the subcommittee concludes that the exposure analysis is lacking in several respects. The DPR report fails to address several exposure scenarios, including exposures of residents living near fumigated fields and increased exposures of residents and workers resulting from methyl bromide treatment of several agricultural fields simultaneously or consecutively. In addition, the subcommittee concludes that there is considerable uncertainty concerning the analytical recovery methods used in the exposure assessment studies. Much of the data presented by DPR is based on single measurements, and no discussion of variability or uncertainty in the measurements is provided. The DPR report also fails to discuss the representativeness of the measurements to the actual exposures experienced by worker or residential populations. DPR makes numerous assumptions regarding durations and levels of exposures, which the subcommittee believes are not explained in sufficient detail to establish whether the assumptions are valid.

Recommendations

- Further data collection and analysis are necessary to accurately assess worker and residential exposures to methyl bromide.
- Improvement is needed in the collection of field data used by DPR to

assess worker exposure, particularly with regard to the analytical methods used to detect methyl bromide in ambient air and atmospheric conditions during sampling.
- Further work is needed to determine the best recovery method for methyl bromide and how field conditions affect the recovery of methyl bromide from air samples.
- Air sampling should be conducted for residents living near fumigated fields; these nonoccupational exposures are unquantified at present.
- DPR should reevaluate all existing exposure data for variability and uncertainty.

Risk Characterization

DPR characterized the risks associated with exposure to methyl bromide by using a margin-of-exposure (MOE) approach. DPR compared the human equivalent NOAEL determined from the available animal toxicity data with the anticipated or measured exposures of agricultural workers and residents located near fumigated fields and those entering fumigated homes. The subcommittee found the MOE approach to be generally acceptable for determining which workers and residents are likely to be exposed to potentially harmful concentrations of methyl bromide. However, the subcommittee believes that DPR did not conduct a complete risk assessment, because there was no quantification of the populations of workers that are likely to be exposed or the number of residents living near fields or entering houses.

The subcommittee found DPR's use of an MOE to be helpful for estimating risks to some populations, particularly workers. However, the subcommittee has concerns about DPR's use of these MOEs for protecting nonworkers, particularly people living near fumigated fields. The DPR document does not indicate how the MOEs are to be used to determine the protectiveness of the buffer zones specified in the application permits. The document also fails to characterize certain potentially sensitive populations, such as children in schools or living near fumigated fields, although the proposed regulations address the exposure of children by restricting the application times near schools. The subcommittee concludes that the uncertainties addressed by DPR in the report, including extrapolating from LOAELs to NOAELs and from animals to humans, although important, are only part of the uncertainties that need to be addressed in the risk characterization document. The subcommittee finds that DPR's use of a factor of 10 to account for intraspecies variation and a factor of 10 for differences in animal and human toxicity, as well as its use of a benchmark MOE of 100, is consistent with generally accepted risk management practices. The subcommittee concluded that an additional safety factor for infants and children was not necessary, because the NOAELs were adequately conservative.

Recommendations

- DPR should quantify the number and distribution of workers and residents potentially exposed to methyl bromide.
- Buffer zones should be derived based on reasonable worst-case exposure scenarios.
- DPR should be more explicit in linking its MOE analysis to the development of regulatory levels and should indicate how its regulatory goals will be met by its risk characterization.

1

Introduction

BACKGROUND

Methyl bromide is a gaseous fumigant that kills insects, rodents, nematodes, weeds, and organisms that cause plant diseases. It is used for pest control in structures such as warehouses, ships, freight cars, and homes, in preplant treatment of soil, and in post-harvest treatment of commodities. Between 1993 and 1997, 14 to17 million pounds of methyl bromide were used annually in California. Methyl bromide is released into the air during and after its use, and therefore, inhalation exposure to agricultural workers and the general population is of considerable concern. The primary health effect of acute methyl bromide exposure is neurotoxicity.

Methyl bromide is a Class I ozone depleter and, as such, it is regulated by the Clean Air Act and the United Nations Montreal Protocol. It is scheduled to be phased out of use in the United States by 2005. Because of its toxicity, several federal agencies have established inhalation exposure levels for methyl bromide. The U.S. Environmental Protection Agency (EPA) reference concentration is 5×10^{-3} milligrams per cubic meter (mg/m^3) (1.3 parts per billion). The Agency for Toxic Substances and Disease Registry has minimum risk levels of 50, 50 and 5 ppb for acute, intermediate, and chronic exposure scenarios, respectively. The Occupational Safety and Health Administration has an 8-hr time-weighted average permissible exposure limit (PEL) of 20 parts per million (ppm), whereas California's PEL is 5 ppm, with a ceiling of 20 ppm (Title 8, California Code of Regulations 1998, Section 5155). For structural fumigation in California, the reentry level is 1 ppm within wall voids (i.e., the cavity inside of walls). The American Conference of Governmental Industrial Hygienists (ACGIH) has a Threshold Limit Value of 1 ppm (3.89 mg/m^3) (ACGIH 1997), and the National Institute of Occupational Safety and Health has an immediately-dangerous-to-life-or-health level of 250 ppm.

CALIFORNIA REGULATIONS

In California, the use of methyl bromide is regulated by permit conditions because it is classified as a restricted material. The California Department of Pesticide Regulation (DPR) develops the permit conditions based on analyses of exposure and toxicity data. The permit conditions specify the minimum mitigation measures that must be used when applying methyl bromide. Permits for methyl bromide use at a specific site and time are issued by local county agricultural commissioners.

In 1992, DPR conducted a preliminary risk assessment on methyl bromide to address acute inhalation exposures of residents reentering fumigated homes. Based on that risk assessment, permit conditions were developed to reduce acute exposures of workers and residents living near fumigated fields. These changes included promulgation of emergency regulations by DPR to require a longer aeration period following fumigation and lowering the reentry level from 5 ppm to 1 ppm in the wall voids. DPR further required that Fact Sheets explaining the potential human hazards of methyl bromide fumigation be distributed to those potentially exposed.

In 1999, DPR conducted another risk assessment that reevaluated the 1992 acute exposure assessment and also considered subchronic and chronic inhalation exposures to methyl bromide from all uses. This revised risk assessment, which incorporates new health effects studies, additional air monitoring data, and newly refined computer models for estimating methyl bromide emissions, is intended to assist DPR in establishing new regulations for permitting the use of methyl bromide (Title 3, California Code of Regulations).

The proposed regulations are designed to enhance protection for children in schools, establish minimum buffer zones around application sites, and set new limits on work hours for fumigation employees (Title 3, California Code of Regulations). In addition to specifications for application rates and depths, tarpaulin thickness, field size, application timing, and duration of fumigation, the proposed regulations include the following specifications: (1) permit applicants must submit a work plan detailing the proposed fumigation to the county agricultural commissioner before methyl bromide use will be approved; (2) neighbors living on sensitive sites (i.e., homes, schools, hospitals, employee housing centers) that are within 300 feet of the outer boundary of the buffer zone must be notified of fumigations, and they also have a right to ask for a second notification 48 hr before the scheduled fumigation; (3) the establishment of minimum buffer zones of 50 feet for workers and 60 feet for residents to replace the suggested minimums of 30 feet (workers) and 100 feet (residents) that are now advisory; and (4) a requirement that injection of methyl bromide be completed 36 hr prior to the start of a school session.

The need for new regulations for permits has been driven by several factors. The primary factor is a California Superior Court decision ordering DPR to adopt more specific regulations on the field fumigation use of methyl bromide by June 2000. In addition, recently conducted toxicological and air-monitoring studies (DPR 1999) have generated new data. The current DPR risk characterization document incorporates these new data for the determination of the risks to workers and the general public from methyl bromide use.

However, for this risk characterization document to be used to support the proposed regulations, DPR is required to "conduct an external scientific peer review of the scientific basis of any new rule" (California Health and Safety Code § 57004). Consequently, DPR requested that the National Research Council conduct a review of its draft risk characterization document and provide a critique addressing the issues identified in the assigned task. This task was assigned to the Committee on Toxicology, which convened the Subcommittee for the Review of the Risk Assessment of Methyl Bromide (see Appendix A for biographical information). In addition, California DPR's risk characterization document has undergone internal review by the California Office of Environmental Health Hazard Assessment (OEHHA), part of the California Environmental Protection Agency. Furthermore, DPR has also requested that the EPA review the document.

THE SUBCOMMITTEE'S TASK

The task given to NRC's subcommittee on methyl bromide states the following: The subcommittee will perform an independent scientific review of the California Environmental Protection Agency's risk assessment document on methyl bromide. The subcommittee will (1) determine whether all relevant data were considered, (2) determine the appropriateness of the critical studies, (3) consider the mode of action of methyl bromide and its implications in risk assessment, and (4) determine the appropriateness of the exposure assessment and mathematical models used. The subcommittee will also identify data gaps and make recommendations for further research relevant to setting exposure limits for methyl bromide.

DPR provided the subcommittee with the draft report to be reviewed *Methyl Bromide: Risk Characterization Document for Inhalation Exposure* (DPR 1999). This report evaluates the toxicological and exposure data on methyl bromide that characterize risks at current exposure levels for field workers and nearby residents. This document comprised the basis for the subcommittee's review. The subcommittee also reviewed much of the primary toxicology literature cited in the risk characterization document, as well as other supporting materials provided by DPR (OEHHA 1999; Seiber 1999).

In addition to DPR's report and supplemental materials, the subcommittee held a public meeting on October 4, 1999, to gather information from DPR and other interested individuals and organizations. At this meeting, formal presentations were made by Paul Gosselin, Lori Lim, and Thomas Thongsinthusak of DPR; Vincent Piccirillo of NPC, Inc.; Bill Walter of the Environmental Working Group; and Amy Kyle of the California Rural Legal Assistance Foundation. A list of materials provided to the subcommittee may be found in Appendix B.

ORGANIZATION OF THE REPORT

The remainder of this report contains the subcommittee's analysis of DPR's risk characterization for methyl bromide. In Chapter 2, the critical toxicological studies and endpoints identified in the DPR document are evaluated. Chapter 3 summarizes DPR's exposure assessment, and the data quality and modeling techniques employed in its assessment are critiqued. Chapter 4 provides a review of DPR's risk assessment, including the adequacy of the toxicological database DPR used for hazard identification, an analysis of the margin-of-exposure data, and appropriateness of uncertainty factors used by DPR. Chapter 5 contains the subcommittee's conclusions about DPR's risk characterization, highlights data gaps, and makes recommendations for future research.

2

Toxicology and Hazard Identification

In this chapter, the National Research Council's subcommittee on methyl bromide reviews the toxicokinetic and toxicological information on methyl bromide as presented in the California Department of Pesticide Regulation's (DPR's) October 1999 draft report *Methyl Bromide: Risk Characterization Document for Inhalation Exposure* (DPR 1999). The information reviewed by the subcommittee is presented in Section III, "Toxicology Profile," and Appendices B and D, of the DPR draft report. In the sections below, the subcommittee comments on DPR's selection of the critical study and toxicological endpoints for acute, subchronic, and chronic exposures. In Table 2-1 below, taken from the DPR risk characterization document (DPR 1999, p. 10), the critical no-observed-adverse-effect levels (NOAELs),[1] toxicity endpoints, and reference concentrations[2] (RfCs) are summarized.

---

[1] The subcommittee has used the term no-observed-adverse-effect level (NOAEL) rather than DPR's term no-observed-effect level (NOEL). NOAEL is defined as an exposure level at which there are no statistically or biologically significant increases in the frequency or severity of adverse effects between the exposed population and its appropriate control. Many organizations use the terms interchangeably.

[2] A reference concentration (RfC) is an estimate of the concentration of a substance that is unlikely to cause noncancer health effects in humans during a lifetime. It is used by DPR as a regulatory value for establishing buffer zones to protect residents from adverse effects of methyl bromide exposure.

TABLE 2-1 Summary of Critical NOAELs Used by DPR

Scenario	Experimental NOAEL (ppm)	Human Equivalent NOAEL (ppm)[a] Adult	Human Equivalent NOAEL (ppm)[a] Child	RfC (ppb)[b]	Effect in Animal Studies	Reference
Acute	40	21	na	210	Developmental toxicity (pregnant rabbit)	Breslin et al. 1990b
	103	45	25		Neurotoxicity (dog)	Newton 1994b
Sub-chronic: 1 wk	20	12	7	120 (adult) 70 (child)	Neurotoxicity (pregnant rabbit)	Sikov et al. 1981
6 wk	0.5 (estimated)	0.2	0.1	2 (adult) 1 (child)	Neurotoxicity (dog)	Newton 1994b
Chronic	0.3 (estimated)	0.2	0.1	2 (adult) 1 (child)	Nasal epithelial hyperplasia (rat)	Reuzel et al. 1987, 1991

[a]The human equivalent NOAEL in parts per million (ppm) is derived from the experimental NOAEL, taking into account the relative breathing rates and exposure durations of animals and humans.

[b]The inhalation reference concentration (RfC), in parts per billion (ppb) is the ratio of the human equivalent NOAEL and an uncertainty factor of 100.

NOTE: na = not applicable.
Source: Adapted from DPR 1999, p. 10.

PHARMACOKINETICS

The absorption, distribution, and, to some extent, the biotransformation and excretion of methyl bromide are reviewed in the DPR report. Following inhalation or ingestion, methyl bromide was absorbed rapidly and distributed to most tissues in the body, based on a ^{14}C label, which can reflect either parent or metabolite (Bond et al. 1985). After 6 hr of nose-only inhalation, rats absorbed 38% to 48% of the methyl bromide concentrations at 1.6 to 170 ppm, but only 27% at 310 ppm (Medinsky et al. 1985); dogs absorbed approximately 40% after 3 hr of exposure to concentrations at 174 to 361 ppb (Raabe 1986, as cited in DPR 1999); and humans absorbed 52% to 55% of concentrations of 18 ppb after 2 hr of exposure (Raabe 1988, as cited in DPR 1999). In rats, a radioactive label (radiolabel) was found in the nasal turbinates, lungs, testes, brain, thymus, and adrenal glands of the rat (Medinsky et al. 1984; Bond et al. 1985; Jaskot et al. 1988). The tissues with the highest radioactivity were lung, liver, and nasal turbinates. Following oral administration in rats, more than 90% of the dose was absorbed (Medinsky et al. 1984). Following a high-dose accidental exposure of a man who died 4 hr later, methyl bromide was detected in all tissues, except the spleen, at an autopsy 1 hr after death (Michalodimitrakis et al. 1997 as cited in DPR 1999).

Methyl bromide is rapidly biotransformed following inhalation (Bond et al. 1985). In rats, more than 75% of the inhaled dose was excreted within 65 hr (Bond et al. 1985). Over 90% of the inhaled radioactivity in rats was associated with metabolites with an elimination half-life from tissues of 1.5 to 8 hr (Bond et al. 1985). About half of the inhaled dose in rats is excreted as exhaled CO_2 with a biphasic half-life of approximately 4 hr and 11 to 17 hr; less than 5% is exhaled as methyl bromide (Medinsky et al. 1985). About 20% of the absorbed radioactivity is excreted in the urine and only about 1% via the feces. The estimated clearance half times of the radiolabels in dogs and humans are about 41 hr and 72 hr, respectively (Raabe 1986, as cited in DPR 1999).

Although the discussion on the absorption, distribution, and excretion of ^{14}C-labeled methyl bromide (based on following the radiolabel) in the DPR reports appears complete, there is only limited discussion of its metabolism. The studies on the excretion of ^{14}C following the inhalation of radiolabeled methyl bromide are consistent with the hypothesis that the methyl group of methyl bromide joins the 1-carbon pool after metabolism. Thus, it is important that the literature on the metabolism of methyl bromide be reviewed in the DPR document. The metabolism of methyl bromide could have implications for its toxicity and subsequent risk assessment, especially for some segments of the population, as noted below.

Following absorption, conjugation with glutathione—a common detoxification mechanism—appears to be the primary metabolic pathway for monohalomethanes, including methyl bromide (Hallier et al. 1990; Peter et al. 1989; Bonnefoi et al. 1991). Methylglutathione is then metabolized to S-methylcysteine by transpeptidases. In turn, S-methylcysteine is metabolized to methanethiol through methylthioacetic acid. Methanethiol is oxidized to formaldehyde and hydrogen sulfide and then to formate and sulfate (Kornburst and Bus 1983).

The toxicity of methyl bromide might result from the direct methylation of cellular macromolecules or the toxic metabolites methanethiol, formaldehyde, or hydrogen sulfide. Exposure to methyl bromide has induced glutathione depletion in some in vivo and in vitro studies. It is possible that in cases of cumulative exposure to chemicals involving glutathione conjugation, less glutathione would be available to detoxify methyl bromide, resulting in more acute effects. Reports on the effect of glutathione depletion or administration of the glutathione precursor (N-acetylcysteine) on the toxicity of methyl bromide are not consistent and vary with the species and endpoint examined (Garnier et al. 1996). The reasons for these inconsistencies are not clear. However, it is evident that the interactions of methyl bromide with glutathione or the metabolites of methyl bromide play a role in its toxicity (Garnier et al. 1996).

A genetically determined polymorphism in glutathione transferase activity has been reported in humans (Hallier et al. 1993). Humans can have at least 14 classes of this enzyme with 11 subunits. Approximately 75% of humans have erythrocytes with a form of this enzyme, which is selective for the conjugation of methyl bromide with glutathione (fast conjugators), but about 25% of the population does not have this enzyme phenotype (slow conjugators). The specific class of glutathione transferase responsible for conjugating methyl bromide and its metabolites is thought to be glutathione transferase theta (GST T1-1) (Garnier et al. 1996). Ethnic differences in the prevalence of the genotype have been reported (Nelson et al. 1995).

In the case of sister chromatid exchanges in blood cells exposed in vitro to methyl bromide, cells from fast conjugators appeared to be protected from the production of sister chromatid exchanges, whereas cells from slow conjugators were not. On the other hand, in a single report involving one fast conjugator and one slow conjugator, the slow conjugator appeared to have considerably fewer and less severe neurological symptoms than did the fast conjugator, indicating that the proximate toxin might have been one of the metabolites of methyl bromide conjugated with glutathione as discussed above (Garnier et al. 1996). These very limited data suggest that the ability to quickly conjugate methyl bromide to glutathione might have profound ramifications on whether or not an exposed individual is likely to experience neurotoxic effects or be more susceptible to genotoxic effects. It is possible that the ability to conjugate methyl bromide with glutathione could influence the dose response for the carcinogenicity, mutagenicity, neurotoxicity, or other toxic effects for certain segments of the population who either possess or lack this polymorphism. Such differences in metabolism and toxic outcomes, if better substantiated, might be a factor in identifiying susceptible subpopulations that should receive special consideration in the risk assessment process (Hallier et al. 1993). In addition, animals used in toxicity studies might not mimic the polymorphisms seen in the human population.

GENOTOXICITY

Methyl bromide is a methylating agent and reacts with cellular macromolecules. As reviewed in Section III.E of the DPR report, methyl bromide is genotoxic in a number of in vitro and in vivo assays. In in vitro studies, it is a direct-acting mutagen in *Salmonella typhimurium* strains TA100 and TA1535, *Escherichia coli* strains Sd-4 and WP2hcr, and in *Saccharomyces cerevisiae* (Simmon et al. 1977, as cited in DPR 1999; Kramers et al. 1985, as cited in DPR 1999; Moriya et al. 1983; NTP 1992; Djalali-Behzad et al. 1982; Mortelmans and Sheperd 1980, as cited in DPR 1999). It produces a dose-dependent induction of sex-linked recessive lethality in *Drosophila melanogaster* (Kramers et al. 1985, as cited in DPR 1999; McGregor 1981, as cited in DPR 1999) and forward mutations in a mouse lymphoma assay (Kramers et al. 1985, as cited in DPR 1999). The DPR report states that, of the gene mutation studies summarized above and specifically cited in the document, only the study that used *Saccharomyces cerevisiae* was considered acceptable by DPR. DPR does not provide an explanation for this assessment, although the subcommittee finds that such an evaluation should be justified. The subcommittee is particularly concerned that DPR has implied that the gene mutation studies performed by National Toxicology Program (NTP) are not satisfactory. If this is the case, DPR should provide a detailed explanation as to why these tests are not acceptable.

DPR evaluated several in vivo assays. In female mice exposed to methyl bromide by inhalation at 100 and 200 ppm for 10 days, an increase of micronuclei was observed (NTP 1992). No increase of micronuclei was seen following intraperitoneal injection of methyl bromide at up to 123 mg/kg (Putnam and Morris 1991, as cited in DPR 1999). Dominant lethal mutations were not observed in rats exposed to methyl bromide by inhalation. A dose-related increase in the frequency of sister chromatid exchanges in bone marrow of female mice exposed to concentrations of methyl bromide at 100 or 200 ppm for 10 days was reported (NTP 1992). However, this was not seen in another study in which mice were exposed to a concentration of methyl bromide at 120 ppm for 12 weeks (NTP 1992). DNA adducts were detected in liver, lung, stomach, and forestomach of rats exposed to high concentrations of methyl bromide at 130 to 260 ppm by inhalation (Gansewendt et al. 1991). Bentley (1994, as cited by DPR 1999) exposed rats for 5 days to concentrations of methyl bromide up to 250 ppm and examined testicular damage. DPR determined that at 250 ppm, methyl bromide was considered positive for genotoxic potential to the DNA of testicular cells, whereas, at lower exposure levels, results were inconclusive. Because of the lack of raw data concerning this study, both DPR and the subcommittee consider this study to be unacceptable.

Although the available literature on the genotoxicity of methyl bromide appears to have been adequately reviewed in the DPR report, there is no discussion of the human erythrocyte polymorphism (Hallier et al. 1993). In addition, there is no mention of the significance of the genotoxicity of methyl bromide in relation to the potential carcinogenicity of methyl bromide. Because methyl bromide is a direct-acting mutagen, especially in in vitro systems, some discussion for its lack of correlation with carcinogenicity should be presented in the DPR report.

ACUTE TOXICITY

In Section III.B of the DPR report, four rat, one mouse, two dog and one guinea pig inhalation toxicity studies were examined for acute effects, as were one rat and one rabbit inhalation developmental studies, and one dog oral study. In addition, human effects from dermal exposures were analyzed. In general, dogs appear to be the most sensitive species of laboratory animals studied. The study by Newton (1994b) in which dogs were exposed via inhalation for 7 hr/day for 34 days over 6 weeks, was considered by DPR as a critical study for neurotoxicity endpoints. The NOAEL and lowest-observed-adverse-effect level (LOAEL) were 103 and 158 ppm, respectively, based on no recorded adverse effects until 9 days of exposure at 103 ppm, and decreased activity on the second day of exposure at 158 ppm, and brain lesions following 6 days of exposure at 158 ppm. This study appears to have been appropriate for selecting a NOAEL because the neurotoxicity endpoint is highly relevant to humans, and the critical signs of neurotoxicity (decreased activity) seen at the LOAEL were noted on the second day of exposure, and not following multiple days of exposure. In addition, most of the other studies would have identified higher NOAELs.

It should be noted that the DPR document is quite confusing with respect to its discussion of the acute critical studies. There is substantial discussion of acute effects under the section on acute toxicity, but the critical study (Newton 1994b) is discussed under the section on subchronic toxicity, probably because this was also the critical study for the subchronic NOAEL. The subcommittee recommends that DPR revise its acute toxicity section to include a discussion of these observations so that there is greater clarity on the endpoints selected for the critical study for the acute NOAEL. In addition, the observation of toxicity after the single dose was a functional one, that is, decreased activity, which is likely to be a more sensitive manifestation of toxicity than pathological lesions. A similar NOAEL was noted in a rat inhalation study with a single 6-hr exposure (Driscoll and Hurley 1993); however, the

LOAEL was considerably higher (350 ppm, based on deficits in a functional observational battery, which is also a functional endpoint similar to that of the critical dog study). Further discussion of the critical dog study (Newton 1994b) is found in the neurotoxicity section of this chapter.

The greater sensitivity of the dog compared with the rat makes the dog study the appropriate critical study for regulatory purposes. Also, inhalation exposure is more likely for acute effects than is oral exposure, and therefore, an inhalation study is more appropriate to designate as the critical study than an oral study. Human dermal exposure data were too limited to consider for regulatory purposes. Lower NOAELs and LOAELs were observed in some in vivo and in vitro studies (Honma et al. 1987 and 1991); however, a critical analysis of these data by DPR revealed a number of inconsistencies in the data sets, rendering them unsuitable as critical studies. The subcommittee concurs with DPR that the Honma et al. studies (1987, 1991) are not appropriate for regulatory purposes because these studies examined neurochemical endpoints that are not necessarily indicative of toxicity.

In addition to neurotoxicity, DPR also considered developmental toxicity for its acute exposure risk assessment, based on the assumption that only a single exposure at a critical time is necessary for the induction of a developmental effect. This is a well-accepted principle in developmental toxicology that has been incorporated into the U.S. Environmental Protection Agency's (EPA's) *Guidelines for Developmental Toxicity Risk Assessment* (1991). A developmental toxicity study in rabbits was considered by DPR as the critical study for developmental endpoints (Breslin et al. 1990b). The LOAEL and NOAEL were 80 and 40 ppm, respectively, with gallbladder agenesis, fused sternebrae, and reduced fetal weights observed at 80 ppm. This study was considered appropriate by the subcommittee for selection of the NOAEL for acute toxicity for women of childbearing age in the workforce and in the general population. This study is discussed in more detail in the developmental toxicity section.

SUBCHRONIC TOXICITY

A large number of studies were available for analysis by DPR for subchronic toxicity involving several species and dosing protocols. Eight rat, three mouse, seven rabbit, two dog, one guinea pig, and two monkey inhalation studies, and four rat oral studies were considered by DPR. As was true for acute toxicity, it is most appropriate to consider inhalation exposure because this route most accurately reflects the primary route of repeated exposures in humans. Because there is a convincing database on neurotoxic effects stemming from human exposures to methyl bromide, the subcommittee concludes that it is most appropriate to consider neurotoxicity as the critical endpoint in animal studies used for risk characterization of subchronic toxicity.

DPR selected two critical studies: a dog study of exposures of 7 hr/day for 34 days over 6 weeks (Newton 1994b), and a rabbit developmental study with exposures of 7 hr/day for gestation days 1 to 15 (Sikov et al. 1981). As with the acute toxicity considerations, nonrodents appear to be more sensitive to methyl-bromide-induced toxicity than are rodents, so selection of nonrodent studies is appropriate. The dog is particularly sensitive, and in the Newton (1994b) study, exposures of dogs provided a LOAEL of 5 ppm. No NOAEL was identified. The LOAEL was based on decreased responsiveness, that is, listlessness and quiescence, and also on the decreased spleen weight observed in female dogs at the end of week 6 of the exposure. The responsiveness, although a subjective observation, is a functional one, and indicates that this LOAEL was based on a highly sensitive endpoint. However, decreased responsiveness is a somewhat equivocal endpoint, in that it apparently was not part of the standardized test protocol, but was an additional observation volunteered by a veterinarian who observed the two female dogs (out of a total of four female and four male dogs). The observation was not dramatic, but because the veterinarian, who would be trained to notice abnormal behavior in dogs, felt that this was worthy of special note, the subcommittee must assume that the two dogs were, indeed, showing aberrant behavior, and that this is logically attributed to methyl bromide. Although there were only two non-responsive dogs, which is clearly a low number of observations, they represented half of this experimental group, suggesting that it was a significant observation. In addition, many of the other studies indicated reduced activity as one of the primary signs resulting from methyl bromide exposure, so the observations of reduced responsiveness in these two dogs are consistent with the overt signs of methyl bromide neurotoxicity observed in those other studies. Therefore, the subcommittee concurs with DPR's evaluation that these observations assessed a toxic response, that this dosage group constituted the LOAEL, and that this 6-week study should be the critical study for the risk assessment of subchronic toxicity. Additional details on the Newton (1994b) study are described in the Neurotoxicity section.

The rabbit developmental study (Sikov et al. 1981) was used as a 1-week subchronic critical study. This study identified a NOAEL and LOAEL of 20 and 70 ppm, respectively, based on convulsions and paresis in the dams; and it demonstrated a steep dose-response curve, based particularly on severe signs of toxicity. A rat study identified a lower NOAEL and LOAEL than the rabbit study, in which rats were exposed to concentrations of methyl bromide at 3 and 30 ppm for 6 hr/day, 5 days/wk for 132 to 145 days (American

Biogenics Corporation 1986). Thus, the LOAEL seen in rats exposed to 30 ppm for 132 to 145 days suggests that this species is less sensitive to methyl bromide than rabbits, which had a NOAEL of 20 ppm for a 7-day exposure. Another rat study also identified a lower NOAEL based on biochemical measures (i.e., decreased brain monoamine levels) (Honma et al. 1982); however, because it is unclear whether these biochemical changes genuinely reflect or correspond with any functional changes, and because of the concerns of design and consistency in the results from the series of Honma et al. studies, this is not a suitable study to select as a critical study. Therefore, the subcommittee concurs with the selection of studies by DPR as the critical studies for subchronic toxicity.

CHRONIC INHALATION AND ONCOGENICITY

Two chronic inhalation studies were reviewed by DPR for the assessment of the chronic toxicity and oncogenicity of methyl bromide (Section III.D). The first study (Reuzel et al. 1987, 1991) involved exposure of male and female Wistar rats to methyl bromide concentrations at 0, 3, 30, or 90 ppm for 6 hr/day, 5 days/wk, for 29 months. Each exposure group comprised 90 males and 90 females with interim sacrifices of 10 rats/sex/group at 13, 52, and 104 weeks. Body weights, clinical signs, hematology, biochemistry, and gross and microscopic effects were examined at these times. Exposure to 90 ppm was clearly toxic with early mortalities reported (not statistically significant at the terminal sacrifice); body weights of both sexes in this exposure group were significantly lower than those of the respective control groups throughout most of the study. At terminal sacrifice, effects on the heart were apparent in the 90-ppm exposure group. Statistically significant higher incidences of heart lesions in this group included cartilaginous metaplasia (males only), moderate to severe myocardial degeneration (females only), and thrombi (males and females). Myocardial degeneration also occurred in aged control rats. Therefore, when total incidences of myocardial degeneration were considered, incidences in the control and 90-ppm groups were similar for both sexes. At the 29-month sacrifice, the 3-ppm concentration was a NOAEL for endpoints of body weight and absolute and relative brain weight. The subcommittee noted that the absolute brain weights were significantly reduced for both sexes in the 3- and 30-ppm groups, but did not consider the reductions biologically significant (98% of control values for both sexes in the 3-ppm group, and 94% and 96% of control values for males and females, respectively, in the 30-ppm group), especially in the absence of histological correlates. DPR reevaluated the absolute brain-weight data by combining incidences at 29 months with data from animals that died during the study. With the reevaluation they found that the NOAEL remained 3 ppm for the reduction in absolute brain weight.

Basal cell hyperplasia of the olfactory epithelium was present in both males and females in a dose-related manner at the 29-month terminal sacrifice, but not at the other time points. Incidences were statistically significant in the 3-ppm group at the 29-month terminal sacrifice when total incidences were considered (13 of 48 and 19 of 58 in males and female subgroups, respectively, compared with 4 of 46 and 9 of 58 in the respective control subgroups). In the 3-ppm group, the majority of lesions were characterized as "very slight"; the severity of these lesions became greater ("slight" to "moderate") in the higher exposure groups. These lesions were not present in either males or females in the 3-ppm group at the 52-week interim sacrifice and were not significantly elevated over those of the respective control groups at the 104-week interim sacrifice. But these lesions were present in the female control group at the 104-week interim sacrifice at an incidence (4 of 10, 40%) similar to that in the female 3-ppm group at terminal sacrifice (19 of 48, 40%). At terminal sacrifice, the incidence of total olfactory lesions in males in the 3-ppm group was 13 of 48 (27%) compared with 4 of 46 (9%) in the male control group.

The subcommittee made the following observations regarding the nasal lesions: (1) they increased in control rats in an age-dependent manner from 12 to 29 months; (2) all but one of the lesions were classified as slight or very slight in the 3-ppm group; and (3) one moderate lesion of the nasal mucosa was observed in controls at the 24-month observation (accompanied by a 40% incidence of total lesions in control females). The incidence in the control males at 24 months was 3 of 10 (30%). Therefore, the effect in the 3-ppm group at the 29-month terminal sacrifice, although dose-related and statistically significant, must be considered slight or equivocal. This study was well conducted, used a relevant route of administration, used adequate numbers of rats of both sexes, and examined all relevant endpoints of methyl bromide toxicity. In addition, the same critical endpoint of more pronounced lesions, was observed in rats exposed for shorter periods of time at higher concentrations (Eustis et al. 1988; Hurtt et al. 1988). Therefore, the subcommittee concurs that this study should be the critical study for the chronic risk assessment. A separate chronic RfC for children based on this study is not necessary because the endpoint is not applicable or relevant to children. The study shows that the endpoint occurs in aged rats; children exposed throughout their childhood will not show this particular endpoint.

The second chronic inhalation study reviewed by DPR was conducted by the National Toxicology Program (NTP 1992). This chronic study with B5C3F1 mice included neurobehavioral evaluations at 3-month intervals and was an equally suitable study for derivation of the RfC. In this study, each group of 70 male and 70 female mice was exposed to concentrations of methyl bromide at 0, 10, 33, or 100 ppm for two years—the typical duration of chronic studies. Sixteen mice (eight males and eight females) per group were used for neurotoxicity testing only. Interim sacrifices of 10 mice/sex/group took place at 6 and 15 months. The exposure to 100 ppm was discontinued after 20 weeks because of neurotoxicity and early mortalities. This study identified the same organ and tissue endpoints as the Reuzel et al. (1987, 1991) study, the nose, heart, and brain, and also included an endpoint for affected bone. Aside from increased mortality in the 100-ppm dose group, statistically significant LOAELs and NOAELs for target organ effects were cerebellar and cerebral degeneration, 100 and 33 ppm; myocardial degeneration and chronic cardiopathy, 100 ppm and 33 ppm; sternal dysplasia, 100 ppm and 33 ppm (increased but not statistically significant for either males or females over controls in the 33-ppm group); and olfactory metaplasia or necrosis, 100 ppm and 33 ppm. It was noted that, similar to results observed in rats at the 3-ppm exposure group in the Reuzel et al. (1987, 1991) study, no olfactory lesions were present in mice at the end of 24 months.

DPR identified a LOAEL of 10 ppm for neurotoxicity in the NTP (1992) study based on statistically decreased locomotor activity at 6 and 12 months. The subcommittee disagrees with DPR that the LOAEL for neurotoxicity is 10 ppm. A statistically significant decrease occurred at only 1 of 8 time periods for each sex (6 months for males and 12 months for females) and these decreases were offset by random nonstatistically significant increases over control values at other test times. The authors of the NTP study found "no consistent neurobehavioral differences in animals from the two lower dose groups" (10 and 33 ppm). Therefore, the LOAEL for neurotoxicity is 100 ppm. The NTP Peer Review Panel concurred with the findings of the authors of the NTP study with one member advising caution in the interpretation of the behavioral and functional neurotoxicity results. The Peer Review Panel comments are incorporated into the NTP report.

DPR also reviewed the chronic inhalation study of Gotoh et al. (1994, as cited in DPR 1999). This study was not considered acceptable by either DPR or the subcommittee because the study was reported in summary form and individual data were not available for evaluation. The two-generation reproduction study by American Biogenics Corporation (1986) can also be considered when evaluating chronic toxicity. However, the NOAEL and LOAEL of 3 ppm and 30 ppm, respectively, for reduced growth of neonatal rats were higher than in the Reuzel et al. (1987, 1991) study.

In addition to the two chronic inhalation studies, DPR reviewed four dietary studies. One was a study in which rats were administered encapsulated methyl bromide for 2 years (Mertens 1997, as cited in DPR 1999), another was a study in which rats were administered fumigated feed for 2 years (Mitsumori et al. 1990), and the other two were studies in which beagle dogs were administered fumigated feed for 1 year (Rosenblum et al. 1960; Newton 1996, as cited in DPR 1999). Although the oral exposure route is not the most appropriate for deriving an inhalation RfC, such studies can be used to identify target organs and potential carcinogenic effects. The studies in which animals were administered fumigated feed suffered from various defects, including lack of analytical determination of methyl bromide concentrations. DPR correctly considered three of these studies unacceptable or supplemental (Mitsumori et al. 1990; Rosenblum et al. 1960; and Newton 1996, as cited in DPR 1999). In the Mertens (1997) study (as cited in DPR 1999), which DPR considered acceptable, nominal methyl bromide concentrations at 0 (basal diet), 0 (placebo microcapsules), 0.5, 2.5, 50, or 250 ppm were administered in the feed to rats for 2 years. DPR estimated a conservative LOAEL in this study of 0.5 ppm based on splenomegaly in male rats (0 ppm, basal diet: 2 of 50; 0 ppm, placebo: 2 of 50; 0.5 ppm: 7 of 50; 2.5 ppm: 10 of 50; 50 ppm: 11 of 50, and 250 ppm: 3 of 50). The subcommittee reviewed the Mertens (1997) study (as cited in DPR 1999) and, based on the absence of (1) a clear dose-response relationship for splenomegaly, (2) histological correlates in the spleen, and (3) effects on hematology and clinical chemistry parameters, disagrees with DPR's assessment. Based on early effects on body weight, the subcommittee believes the LOAEL is 250 ppm of methyl bromide in the feed and the NOAEL is 50 ppm.

DPR also evaluated the carcinogenicity of methyl bromide. It concluded that although methyl bromide is genotoxic without metabolic activation and has been shown to alkylate DNA in different organs in in vivo studies, there is no clear evidence of oncogenicity under the experimental conditions used in the chronic inhalation studies with rats and mice (Reuzel et al. 1987, 1991; NTP 1992). The subcommittee reviewed the chronic studies for oncogenicity in male and female rats and mice and agrees with DPR's conclusion. The chronic oral study with rats (Mertens 1997, as cited in DPR 1999) also was negative for oncogenicity, supporting the conclusion drawn from the two chronic inhalation studies.

REPRODUCTIVE TOXICITY

Two reproductive toxicity studies were evaluated in Section III.F of the DPR report. Both studies used rats as the experimental animals; one was an inhalation study (American Biogenics Corporation 1986; Hardisty 1992, as cited in DPR 1999; Busey 1993, as cited in DPR, 1999), and the other was an oral study (Kaneda et al. 1993). The inhalation study was a two-generation study designed to investigate the reproductive toxicity of methyl bromide at concentrations of 3, 30, and 90 ppm for 6 hr/day, 5 days/wk. Exposure to these concentrations did not affect the fertility indices of the F0 animals for two separate mating trials, nor did it affect the fertility index of the F1 animals for the first mating trial. However, for the second mating trial of the F1 animals, the fertility indices (number of pregnancies per number of copulations) in the 30- and 90-ppm groups were reduced to 66.7% and 68.2% from 85% to 100% in the other mating trials. In the DPR report, the fertility index for the 30 ppm group is correctly given as 66.7% (based on our calculation from the original data). It is incorrectly given as 70% in Table 8 of the original study (American Biogenics Corporation, 1986). According to the DPR document, this reduction approached statistical significance in both cases (0.056 and 0.066, respectively); however, the DPR report does not state what statistical test was used. The subcommittee calculated p-values of 0.12 (Pearson's chi square) and 0.06 (Mantel-Haenszel test for linear association) for the effect of treatment on whether or not a copulation resulted in pregnancy.

The reduced-fertility indices in the second mating trial of the F1 parents were considered evidence of an effect of methyl bromide on fertility by DPR, although not by the authors of the study, who stated only that there was no significant effect of treatment on the fertility index. Although of borderline statistical significance, the fact that the reduction in fertility occurred in both of the higher-dose groups and only after 90 to 105 days of exposure to methyl bromide (as compared with 40 to 55 days for the first mating) suggests that this was a real effect on fertility. However, this might represent a developmental effect on the reproductive system rather than a reproductive effect because it only occurred in the F1 generation, suggesting that gestational or early postnatal exposure is required to manifest the effect. At the time of sacrifice of the F1 parents at about 250 days, ovary weights were not affected by methyl bromide exposure, but there was a downward trend in absolute testicular plus epididymal weights and a significant reduction in the relative testicular plus epididymal weight in the 30-ppm group. Reproductive organ histology performed on the control and 90-ppm F1 parental animals was reported to be normal. In conclusion, the data suggest that exposure to methyl bromide concentrations at 30 and 90 ppm might be associated with reproductive toxicity; however, the effects were of borderline statistical significance, and the study design does not allow the subcommittee to sort out whether the putative effect was a reproductive or developmental effect of methyl bromide. Therefore, the subcommittee concludes that although 3 ppm was clearly a NOAEL for reproductive toxicity, it is not clear that 30 or 90 ppm were LOAELs.

In addition to the effects on fertility, effects on offspring body weights, organ weights, and brain weight and dimensions are discussed in Section III.F.1 of the DPR report. These should also be mentioned in Section III.G under developmental effects. The significant, dose-dependent reductions in the body weights during lactation of the offspring of all four mating trials might be due to gestational or lactational exposure to methyl bromide. The latter is suggested by the fact that pup weights were not decreased consistently on post-natal day 0 or day 4, but were decreased on postnatal days 14, 21, and 28 in the 30- and 90-ppm groups. In the F2a and F2b progeny, pup weights were also significantly reduced on postnatal days 0, 4, and 7, suggesting that at least some of the effect on pup weight was due to gestational exposure. Methyl bromide exposure of the dams was temporarily halted from gestational day 21 until postnatal day 4, and exposure of the pups was not begun until weaning at postnatal day 28 in all four trials. Therefore, from post-natal days 4 through 28 the pups would have been exposed to methyl bromide only via the breast milk. A literature search revealed no data on the excretion of methyl bromide in breast milk; however, as a lipophilic molecule, it might well be excreted in breast milk. (Data are available that suggest that bromide, that is, sodium bromide, will be excreted in milk [Disse et al. 1996].) The body weight differences did not remain significant in the F1b animals during adulthood. None of the F1a, F2a, or F2b animals was followed into adulthood. The body weights of the F0 females were not affected by methyl bromide exposure beginning in early adulthood (62 days old), but the 90-ppm F0 males showed decreased body weights compared with controls from exposure week 3 (around 80 days old) until the final sacrifice (about 250 days old).

The offspring of the F1b and F2b mating trials and the F0 adult males showed dose-related reductions in brain weights, which were significant for the 90-ppm F0 males, the 90-ppm F1b males and females that were sacrificed as adults, and the 90-ppm F2b females sacrificed at 28 days. Unfortunately, brain weights and other organ weights were not measured in the F1a or F2a offspring. Brain weights were not significantly reduced in the subset of F1b offspring that were sacrificed at 28 days ($p = 0.17$, analysis of variance [ANOVA] performed by the subcommittee; see Table 2-2). Cerebral cortex widths were measured in the 0- and 90-ppm groups of the adult F0 and F1b animals, and were significantly reduced only in the 90-ppm F1b animals. Absolute organ weights (heart, kidney, liver), but not organ weights adjusted for body weight, were significantly reduced in the 90-ppm F2b female progeny and nonsignificantly reduced in the male F2b progeny (kidney, liver, testis). Taken together, the cortex width data, the body and organ weight data, and the fertility data suggest that the developing rat is more sensitive to methyl bromide toxicity than adults are. The subcommittee concurs with DPR that the

TABLE 2-2 Brain-Weight Summary for F1b Weanlings Sacrificed at 28 Days, by Exposure Level

F1b weanlings	0 ppm	3 ppm	30 ppm	90 ppm
Males[a]	1.51 ± 0.14 g	1.52 ± 0.11 g	1.47 ± 0.10 g	1.50 ± 0.09 g
Females[a]	1.48 ± 0.09 g	1.42 ± 0.12 g	1.41 ± 0.12 g	1.45 ± 0.06 g

[a]No statistically significant differences by ANOVA.

developmental NOAEL for this study is 3 ppm, based on significant body-weight reductions in the offspring of the 30- and 90-ppm groups.

The second reproductive study (Kaneda et al. 1993) discussed in Section III.F.2 of the DPR report was designed to test the effects of methyl-bromide-fumigated feeds. Because the feeds were allowed to aerate for 21 days after fumigation, the doses of methyl bromide were quite low (maximally 200 ng/kg/day, actual doses not determined) compared with the bromine doses (about 110 to 730 µg/kg/day). No effects on mating index or fertility index were observed. The concluding sentence of this paragraph, is confusing "[s]ince the actual methyl bromide concentration in each dose is not known, it is not possible to determine whether the effects were due to bromine or methyl bromide" (DPR 1999, p. 72). It would be more accurate to say that the study cannot be used to establish NOAELs for reproductive effects for methyl bromide because the actual doses of methyl bromide were not determined. Moreover, the inhalation route of exposure is more relevant to humans.

In addition to the two reproductive studies reviewed by DPR, there is another study (Sikov et al. 1981), reviewed by DPR as a developmental toxicity study, which provides some additional information about the reproductive effects of methyl bromide. The report describes a study of pregestational and gestational inhalation exposure of female Wistar rats to nominal concentrations of methyl bromide at 0, 20, and 70 ppm for 7 hr/day, 5 days/wk. The pregestational exposures occurred for 3 weeks before mating with untreated males. No significant effects were observed on fertility rates (92% to 100% in the exposed groups compared with 98% in the controls), on corpora lutea per dam, implants per litter, or implants per corpus luteum. These data are consistent with the results of the inhalation study discussed above (American Biogenics Corporation 1986; Hardisty 1992, as cited in DPR 1999; Busey 1993, as cited in DPR 1999) that found no effects on fertility indices of the F0 generation exposed for up to 105 days before mating.

The subcommittee found DPR's discussion of the reproductive studies of methyl bromide to be somewhat contradictory. In Section II of the DPR report, DPR found that methyl bromide is a direct reproductive toxicant; however, in Section IV, in its discussion of the risk assessment of methyl bromide, DPR suggested that the reduced fertility seen in the American Biogenics Corporation (1986) study might be the result of developmental effects on the reproductive system, resulting in altered reproductive function in adult life. The subcommittee concurs with DPR's latter conclusion that, taken together, the results of the Sikov et al. (1981) and American Biogenics Corporation (1986) studies suggest that methyl bromide might affect the development of the reproductive system, but the subcommittee does not agree that the studies support DPR's conclusion that methyl bromide is a direct reproductive toxicant in adult animals.

DEVELOPMENTAL TOXICITY

Four inhalation and two oral exposure studies in two species were reviewed by DPR for the assessment of the developmental toxicity of methyl bromide (Section III.G). The developmental aspects of the rat inhalation study (American Biogenics Corporation 1986; Hardisty 1992, as cited in DPR 1999; Busey 1993, as cited in DPR 1999) discussed in Section III.F.1 of the DPR report should also have been discussed in Section III.G.

The inhalation study described above (Sikov et al. 1981) also examined developmental endpoints in rats. This study examined the effects of exposure to methyl bromide at 0, 20, and 70 ppm before or during gestation. Minimal maternal toxicity, in the form of reduced gestational body weights on some gestational days in the dams exposed to 70 ppm during gestation, was observed. The investigators also observed higher, although not statistically significant, rates of various ossification defects in the rat fetuses whose dams were exposed to methyl bromide, compared with those who were not. For the most part, these did not show a clear dose-related pattern. However, for the supraoccipital, interparietal, and parietal bones of the skull, the percent of litters and the percent of fetuses that displayed ossification defects were higher in the 20- and 70-ppm gestationally exposed groups than in the controls or in the groups with only pregestational exposure (see Table 2-3). Although such ossification defects are often considered to represent variants when they are consistently elevated in the experimental groups in a manner that suggests dose-dependency, there should be concern about subtle developmental effects (EPA 1991; Sikov et al. 1981). The skull ossification defects, as well as the total skeletal anomalies and total ossification defects listed in Table 2-3, fulfill the consistency criterion (all gestationally exposed groups have higher rates than the controls), but the dose-dependency criterion is only fulfilled for supraoccipital ossification and total ossification defects. In summary, these data suggest that methyl bromide might be a developmental toxicant at doses as low as 20 ppm; however, they do not unequivocally establish it as a developmental toxicant. Therefore, this study should not be used to set a developmental NOAEL.

The same paper (Sikov et al. 1981) also described an inhalation study in rabbits using similar exposure regimens that were planned to continue for gestation days 1 to 24, without pregestational exposure. All of the does in the 70-ppm group manifested severe toxicity, and all but one of them died, even though the exposures were terminated early, on gestation day 15. The NOAEL for maternal toxicity was 20 ppm. No adverse effects were observed in the offspring of the 20-ppm group or of the one survivor in the 70-ppm group. However, because the exposures were terminated early in all groups, this study was not considered to be a valid study of developmental toxicity by DPR. Nonetheless, it provides evidence that gestational exposure to 20 ppm methyl bromide from gestation days 1 to 15 does not cause developmental toxicity in rabbits.

TABLE 2-3 Summary of Skull Ossification Defects in Sikov et al. (1981) Rat Substudy

Ossification defect	Exposure levels in ppm (pre-mating/gestation)						
	0/0	20/0	70/0	0/20	0/70	20/20	70/70
Supraoccipital	2.7[a]	11.8	0	6.5	19.4	5.3	19.4
	0.5[b]	1.9	<0.4	1.1	4.3	0.8	8.7
Interparietal	13.5[a]	14.7	13.9	29.0	19.4	21.1	38.9
	2.9[b]	2.3	2.1	7.0	3.9	3.7	8.7
Parietal	8.1[a]	5.9	11.1	12.9	13.9	28.9	19.4
	1.9[b]	1.4	4.1	4.3	2.6	8.2	4.8
Total skeletal anomalies	0.18[c]			0.30	0.19	0.22	0.26
	1.1[d]			1.4	1.2	1.3	1.7
Total ossification defects	0.03[e]			0.06	0.07	0.07	0.11
	0.24[f]			0.48	0.61	0.68	0.81

[a] Percent litters with defect
[b] Percent pups with defect
[c] Number of skeletal anomalies/fetus
[d] Number of skeletal anomalies/litter
[e] Number of ossification defects/fetus
[f] Number of ossification defects/litter

A preliminary study by Breslin et al. (1990a) examined the effects of inhalation exposure in rabbits to methyl bromide at 0, 10, 30, or 50 ppm (Part 1) and at 0, 50, 70, and 140 ppm (Part 2) for 6 hr/day on gestation days 7 to 19. The exposure regimens used in both Parts 1 and 2 of the study were designed to assess the levels at which maternal toxicity and embryo lethality might occur. No toxicity was observed in the dams or in the offspring in Part 1. In Part 2, the dams exposed to 140 ppm showed severe neurotoxicity with meningeal inflammation and midbrain necrosis, and were sacrificed early on gestation day 17. The dams exposed to 70 ppm exhibited statistically significant decreased body weight at gestation day 16 and decreased weight gain on gestation days 13 to 16 only. Information on maternal weight gain corrected for gravid uterine weight was not given. Therefore, the subcommittee cannot judge whether the effects on maternal weight gain represented maternal or fetal toxicity or both. This study was appropriately considered supplemental by DPR.

A definitive study by Breslin et al. (1990b) exposed rabbits to methyl bromide at 0, 20, 40, or 80 ppm (Part 1) and at 0 or 80 ppm (Part 2) for 6 hr/day from gestation days 7 to 19. Part 2 of the study was designed to determine whether observations made in Part 1 could be replicated. In Part 1, at the 80-ppm dose, 3 of 26 does exhibited clinical signs of neurotoxicity beginning on the last day of exposure. One of these rabbits delivered its litter early, on gestation day 27. Unfortunately, the brains of these animals were not examined. No maternal neurotoxicity was observed in Part 2, but one doe died of undetermined causes in the 80 ppm group. Statistically significant decreases in maternal weights in the 80-ppm group were observed on gestation days 13 and 16 in Part 1, but not at all in Part 2. The two animals with the largest weight losses in Part 1 were also two of the three that displayed neurotoxicity. Statistically significant decreases in maternal weight gains were observed for the interval gestation days 13 to 16 in Part 1 and for the interval gestation days 10 to 13, 7 to 20, and 0 to 28 in Part 2. However, fetal and gravid uterine weights were also significantly decreased in Part 2 and gravid uterine weights were nonsignificantly reduced in the 40- and 80-ppm groups compared with controls in Part 1, suggesting that the reduced maternal weight gain might represent a developmental effect rather than or in addition to a maternal effect. In Section I.C of the DPR report (1999, p. 6), maternal weight gain corrected for gravid uterine weight—a better indicator of strictly maternal toxicity—was described as being unaffected by treatment; however, it does not appear in the results sections of the Breslin et al. (1990b) report. If DPR calculated this parameter from data in the study, the results of the calculation should be included in an appendix.

Fetal malformations were observed primarily in the 80-ppm groups in both

Parts 1 and 2 (Breslin et al. 1990b). The total rate of malformations in Part 1 was 14.5% in the 80-ppm fetuses, compared with 2.1% in the sham-treated control fetuses. In Part 2, a similar rate was observed in the 80-ppm fetuses (14.1%), but a much higher rate was observed in the sham-treated controls (12.3%) compared with Part 1 and compared with the naive controls in Part 2 (5.9%). Malformations included statistically significant increases in gallbladder agenesis and fused sternebrae in Part 1. These defects were observed in fetuses from dams with and without neurotoxicity. A similar incidence of gallbladder agenesis was observed in Part 2, although it did not reach statistical significance with a smaller number (N) of animals.

Although it is not considered to be a major malformation by some experts (Tyl 1991; OEHHA 1993), several arguments favor considering the increased incidence of gallbladder agenesis in the Breslin et al. (1990b) study to be evidence of developmental toxicity of methyl bromide. First, it represents the failure of an entire organ to form. Second, the 8.2% incidence at 80 ppm (affected fetuses/total fetuses) of gallbladder agenesis in Part 1 was statistically significantly elevated over the control incidence of 1.1%. The incidence of 4.3% at 80 ppm in Part 2, although not statistically significant, was nearly 5 times higher than the 0.9% incidence observed in the sham-treated control group. Moreover, the incidence of gallbladder agenesis in the methyl-bromide-treated fetuses is much higher than the 0.09% to 0.19% observed in historically untreated control New Zealand White rabbits from Dow Chemical Company (where the Breslin study was performed), WIL Research Laboratories, the Middle Atlantic Reproduction and Teratology Association, and Stadler et al. (1983) (summarized in Tables 5-8, Appendix B of the DPR report). Third, the same laboratory which performed the Breslin et al. (1990b) study (Dow Chemical Company) has reported gallbladder agenesis as a possible treatment-induced effect of other test compounds (DPR 1999, Appendix B, p. 179).

Kaneda et al. (1998) studied the developmental toxicity of oral methyl bromide exposure in rats (at 0, 3, 10, or 30 mg/kg/day, gestation days 6 to 15) and rabbits (at 0, 1, 3, or 10 mg/kg/day, gestation days 6 to 18). Maternal toxicity in the form of erosion and thickening of the stomach wall and adhesions between stomach and other organs was observed in the rats given 30 mg/kg. Fetuses from that group exhibited statistically nonsignificant increases in microphthalmia and skeletal variations. No maternal effects were noted in the rabbits. In the fetuses, skeletal malformations were observed more frequently in the methyl-bromide-treated groups than in controls, but the increases were not dose-dependent or statistically significant. Because the oral route of exposure to methyl bromide is less significant than the inhalation route for humans, these studies were appropriately considered supplemental information by DPR. To better be able to compare the inhalation with the oral developmental studies, it would be useful to calculate the estimated absorbed doses for these studies.

In conclusion, DPR found that the currently available evidence suggests that methyl bromide might be a developmental toxicant by the inhalation route in two species. In rats, this evidence includes significant, dose-dependent reductions in pup body and brain weights and cerebral cortex widths and a nonsignificant reduction in fertility in gestationally exposed offspring (American Biogenics Corporation 1986; Hardisty 1992, as cited in DPR 1999; Busey 1993, as cited in DPR 1999). As this was a two-generation study, it is not possible to determine the critical period or periods for exposure for these effects. However, the patterns of occurrence are consistent with developmental effects. In a separate rat developmental study, a nonsignificant, but consistent, increase in the incidence of skull ossification defects was observed in gestationally exposed animals (Sikov et al. 1981). In rabbits, the evidence for developmental toxicity includes gallbladder agenesis, reduced fetal weights, and increased frequency of fused sternebrae (Breslin et al. 1990b). Taken alone any of these studies could be considered equivocal; however, taken together, the subcommittee agrees with DPR that they suggest that methyl bromide might be a developmental toxicant.

NEUROTOXICITY

There is ample evidence that neurotoxicity is the most prominent type of toxicity elicited by methyl bromide in humans. Therefore, selection of neurotoxic endpoints for the critical studies in the risk characterization is most consistent with protection of humans from the adverse effects of methyl bromide. Many of the studies did not include neurobehavioral analysis as their endpoints. The critical studies selected for acute and subchronic exposures were based on functional endpoints, and thus appear to reflect the most sensitive indicators measured. The subcommittee finds that a change in behavior (of the dogs in the Newton [1994b] study) is a functional one, which might be more sensitive than a pathological endpoint, because the functional change might be due to a biochemical deficit (or change) even though a histological change has not occurred. As mentioned above, some of the observations in the subchronic study (Newton 1994b) were equivocal, such as the two nonresponsive dogs at the lowest concentration tested; there were low animal numbers in this experiment, not a good dose-response relationship, and the observations were outside the standardized protocol. Nevertheless, the observation of low levels of responsiveness is very consistent with the nervous system depression observed in a number of the animal studies, and therefore is believed to reflect a true neurotoxic response.

One possible exception to DPR's use of the most sensitive functional data available as the critical study would be DPR's rejection of the Honma et al. studies (1987, 1991) on mechanisms for use in the risk characterization. These studies investigated neurochemical endpoints in an attempt to elucidate the mechanisms by which methyl bromide exerts its neurotoxic effects, and implicated primarily the monoaminergic and catecholaminergic systems as potential targets. However, changes in neurochemistry alone do not necessarily indicate toxicity (Overstreet et al. 1974), and these studies did not appear to correlate these neurochemical changes with functional deficits; therefore, the significance of these neurochemical changes cannot be discerned. In addition, there were serious concerns raised by the DPR staff about the design of these experiments because the descriptions of design and methods were overly brief, and it was not clear to the staff whether confounders that could have caused the observed effects had been ruled out. Such issues as contradictions in the data sets obtained or lack of clear dose-response relationships leave these data sets suspect with respect to their suitability for use in a risk characterization. Therefore, the subcommittee concurs with DPR's conclusions that these Honma et al. studies, although suggesting highly sensitive endpoints, are not suitable for use in a risk characterization.

Therefore, neurotoxicity appears to be a prominent form of toxicity elicited by methyl bromide in a variety of laboratory animal studies and is consistent with evidence of neurotoxicity from human studies. Of particular note are studies indicating neurological deficits occurring in those occupationally exposed to methyl bromide at apparently relatively low exposure levels (Anger et al. 1986); these human observations suggest toxic endpoints that need to be considered for a health-protective risk characterization.

SELECTION OF CRITICAL EFFECTS FOR ACUTE TOXICITY

The use of a developmental toxicity study for the assessment of the risks of acute exposure to methyl bromide is a reasonable one, given the principle that a single gestational exposure is sufficient to produce an adverse developmental effect, and in light of the large numbers of women of childbearing age in the workforce. In fact, the EPA Guidelines for Developmental Toxicity Risk Assessment (EPA 1991) state that data from reproductive and developmental toxicity studies should be used in the overall assessment of risk of a compound.

DPR's rationale for using the developmental toxicity studies by Breslin et al. (1990a,b) for assessment of the risk of acute exposure in women of childbearing age is outlined in the document and will be briefly summarized here. First, gallbladder development in the rabbit occurs over 1 to 2 days beginning on gestation day 11.5. Therefore, the assumption that only a single exposure is necessary for the induction of adverse developmental effects is very likely to hold true for this particular effect. Second, the finding of gallbladder agenesis was confirmed when the experiment was repeated. Third, the developmental effects of gallbladder agenesis and fused sternebrae should not be discounted because maternal toxicity occurred in some does. Both defects occurred in offspring of does who did not exhibit neurotoxicity as well as in the offspring of the minority of does who did. In Part 2 of the study, none of the does exhibited neurotoxicity, yet a similar incidence of gallbladder agenesis was observed as in Part 1. In addition, signs of neurotoxicity appeared on gestation day 19, the day 12 of the 13-day exposure. Given that gallbladder development would have been completed 5.5 to 6.5 days before maternal toxicity occurred, it is unlikely that this defect was the result of the maternal toxicity. Moreover, it is not clear that the inconsistent maternal weight changes, which were observed in the 80-ppm groups (significantly lower maternal weight gain compared with controls on some, but not all, gestational days, primarily in Part 2), represented maternal rather than fetal toxicity, because the fetal weights and gravid uterine weights were also significantly lower in the 80-ppm group in Part 2. In addition to DPR's arguments, it should be noted that maternal toxicity has been associated with some developmental abnormalities, including fused sternebrae, but not gallbladder agenesis (Khera 1984; Khera et al. 1989). Moreover, a recent study found no correlation between maternal body-weight change as an indicator of maternal toxicity and various embryo or fetal parameters, including number of anomalies per litter and reduced fetal weight (Chahoud et al. 1999). Finally, another interpretation of the occurrence of maternal and fetal toxicity at the same dose level is that the threshold for toxicity of the test compound is similar in mother and fetus.

The subcommittee considers there to be several counterarguments to using the Breslin et al. (1990b) study to determine the critical NOAEL for acute exposure in women of childbearing age. One, alluded to above, is that the findings of this study, taken on their own, might be considered equivocal evidence for developmental toxicity. The three significant effects—fetal weight decline, fused sternebrae, and gallbladder agenesis—were not statistically consistent between Parts 1 and 2 of the study. Fetal weight was statistically significantly lower in Part 2 in the 80-ppm group, but not in Part 1; gallbladder agenesis was statistically significantly elevated only in Part 1; and skeletal examinations were only performed in Part 1. In addition, fused sternebrae are considered a morphological variation that occurs in 0.27% to 0.92% of untreated control New Zealand White rabbit fetuses (DPR 1999, Tables 5-7, Appendix B). Gallbladder agenesis is a less common developmental abnormality, but it is also considered a variation by some experts. These arguments against the Breslin et al. (1990b) study are weakened by the finding of probable developmental toxicity associated with methyl bromide exposure in another species, the rat, at nonmaternally toxic doses (American Biogenics Corporation 1986; Hardisty 1992, as cited in DPR 1999; Busey 1993, as cited in DPR 1999). Another counterargument is that developmental studies should not be used for acute exposure risk assessment if appropriately performed acute exposure studies exist for the agent in question. This argument ignores the possibility that fetuses might be more sensitive than adults to a given agent and that developmental effects caused by multiday gestational exposures would theoretically be caused by single exposures as well.

Based on the above considerations, DPR's use of the Breslin et al. (1990b) developmental toxicity study to determine the critical NOAEL for acute toxicity for workers and residents appears to be a conservative approach, but one that is justified in the absence of additional data that show that a single exposure at the time of gallbladder development does not cause gallbladder agenesis.

3

Exposure Assessment

BACKGROUND

The National Research Council (NRC) defines exposure to a contaminant as "an event that occurs when there is contact at a boundary between a human and the environment with a contaminant of a specific concentration for an interval of time; the units of exposure are concentration multiplied by time" (NRC 1991). To reliably estimate exposure, environmental monitoring should be conducted to determine contaminant levels, modeling can be used to supplement the monitoring, and potentially exposed populations must be identified and enumerated. Conducting a good exposure assessment requires characterizing the real variability in exposures that are experienced by different groups of people, and different individuals within those groups. In addition, a good exposure assessment integrates an analysis of the uncertainty of the exposure data.

The California Department of Pesticide Regulation's (DPR's) risk characterization document provides exposure estimates for a wide variety of worker and resident exposure scenarios in Sections IV.B, Risk Assessment, and V.C, Risk Characterization, as well as in Appendices F-K. Summary estimates of exposure to methyl bromide, listed in Tables 16-20 of the DPR report, correspond to occupational (Section IV.B.1) and residential (Section IV.B.2) exposure scenarios. The estimates of exposure presented in Tables 16-20 are based on exposure data contained in a report "Estimation of exposure of persons to methyl bromide during and/or after agricultural and non-agricultural uses" by Thongsinthusak et al. (1999) (HS-1659), which is included as Appendix F of the main document.

The exposure information collected in the DPR report came from numerous studies that were conducted for a variety of purposes by several registrants, and therefore were not conducted in a consistent manner nor were they part of a comprehensive and systematic monitoring plan. For instance, DPR points out that many of these studies were not conducted in compliance with Good Laboratory Practices as described in 40 CFR 160 (EPA 1997). Although a variety of analytical techniques were used to determine methyl bromide concentrations in the air samples, these were not reliably tested. In addition, data were collected under different sampling protocols and field conditions (e.g., temperature, relative humidity). For some exposure scenarios, DPR used "default" values due to the lack of specific data on the specific exposure scenarios.

In addition to the limitations described above, DPR acknowledges that the exposure data set is incomplete, as not all potential exposure scenarios are discussed. As stated in the Thongsinthusak et al. (1999) report, "The Department of Pesticide Regulation does not have data to assess all worker exposure scenarios or potential exposure to the public from all methyl bromide applications." However, DPR fails to enumerate what these data gaps are. The lack of a discussion in the DPR report of the limitations of the exposure data set, including the data gaps, undermines the subcommittee's confidence in the data presented by DPR.

The remainder of this chapter addresses the following three aspects of an exposure assessment; (1) the scenarios used to characterize different exposure groups; (2) the quality of data available for characterizing exposures, including the analytical methods used to quantify the air concentrations, and the representativeness of the available air sampling that was conducted; and (3) the modeling used to estimate exposures that were not directly measured. For each of these items, the subcommittee assesses DPR's treatment of the data and its methodology for estimating exposure.

LIKELY EXPOSURE SCENARIOS

The DPR document describes a wide variety of occupational and some residential exposure scenarios. DPR presents valuable information on the uses of methyl bromide in Tables 2 through 5 of Appendix E (pp. 248-250), which provide an understanding of where the most likely exposures might occur. Approximately 95% of the methyl bromide consumed in California is used in soil fumigation, so this mode of use is necessarily a major focus of the analysis. Structural fumigation comprises about 3% to 4% of the methyl bromide use, and commodity fumigation comprises a relatively minor proportion, about 1% to 2%. Based on these use data for methyl bromide, the committee believes that it is important to describe the exposure scenarios within the following categories: (1) occupational; (2) residential, school, and other; and (3) residents returning to fumigated houses. Each of these categories and DPR's coverage of these exposure groups is addressed below.

"Occupational" refers to people who work directly in or around fumigation operations. These individuals are likely to have the most intense exposures and include such labor categories as field applicators (soil fumigators—including pilots, copilots, shovelmen, and workers who remove tarps), structural applicators, and commodity fumigators and aerators. The occupational exposure estimates presented in the DPR report are based on measurements conducted in soil fumigation and commodity fumigation scenarios. The jobs evaluated for exposure and the corresponding estimates of exposure are listed in Tables 16-20 (DPR 1999, pp. 96-106), which include estimates for acute (daily), short-term (7-day), seasonal (90-day), and chronic exposures (annual). A total of 160 exposure categories are listed. Most of the exposure data were measured with personal monitoring devices. The exposure estimates are reported in parts per billion (ppb) and the acute exposure category includes both high and mean values. All other exposure categories are listed as mean values. The 24-hr time-weighted maximal exposures range from a high of 8,458 ppb for sea-container aerators to a low of 0.6 ppb for shallow-shank nontarped bed shovelmen. Numerous job exposures are listed as "n/a," which the table footnote explains as either "not applicable" or "no exposure information available." Unfortunately, it is not clear to the subcommittee which situation applies for a given job category and there is no explanation as to why certain categories of exposure are not applicable to certain jobs.

"Residential, school, and other exposures" refers to people who are exposed to methyl bromide due to its atmospheric transport from the site of direct application. This category specifically includes residents in houses, students in schools, and occupants of buildings near fumigated fields, structures, or near fixed commodity fumigation facilities. This category is expected to contain the most sensitive groups of potentially exposed persons, because it is a cross section of the entire population, and therefore would include the very young and old, as well as other persons that might have heightened sensitivity.

DPR provides no data on exposures to individuals in homes or other buildings near fumigated fields; however, it does provide exposure data on structural fumigations. Gibbons et al. (1996a,b, as cited in DPR 1999) measured methyl bromide concentrations for 24-hr periods in houses located within 50 to 100 feet of fumigated houses. Air sampling in the nonfumigated houses was conducted in rooms closest to the fumigated houses. The measured concentrations range from 0.024 parts per million (ppm) (the limit of detection) to

0.406 ppm. It is unclear from the DPR report how many samples were non-detects. Mean concentration values were 0.024 ppm for nearby houses and 0.060 ppm for "downwind" houses. Downwind is not defined in the DPR report.

Information on exposures to people in residences, schools, and unrelated workplaces near commodity fumigation facilities is based on exposure estimates for workers in those facilities (Haskel 1998a,b). No actual air sampling was conducted to evaluate this nonworker scenario. The assumptions used in this scenario (DPR 1999, Appendix H, page 343) specify that residents are exposed to methyl bromide concentrations at 210 parts per billion (ppb) (24-hr time-weighted average), the maximum permissible exposure level specified in the permit. The subcommittee considers the information provided by DPR insufficient for evaluating the quality of the data used for this assumption and for evaluating the validity of extrapolating from worker exposures to exposures of nearby residents.

"Residents returning to fumigated houses" can be subjected to a wide variety of concentrations, depending on the characteristics of the house and the retention of methyl bromide in spaces in the houses, such as wall voids. This exposure group includes highly-susceptible individuals such as children (NRC 1993), the ill, the elderly, and those with genetic polymorphisms (see Chapter 2). DPR presents exposure measurements from five houses in southern California that were fumigated on a single day followed by 24-hr of active aeration, such as with a fan. (These data are discussed in greater detail below in the section entitled "Exposure of Residents in a Fumigated House".)

In addition to DPR's coverage of the three exposure scenarios above, there are other population groups and exposure scenarios that are never addressed by DPR. For example, DPR never describes or provides data on exposures to children and the elderly, who might be more sensitive to methyl bromide than the worker or general adult populations. Furthermore, DPR never addresses exposure scenarios for residents living near fumigated fields, because these homes were considered to be outside of the permit buffer zone. Therefore, DPR assumed that the maximum concentration to which these individuals could be exposed was 210 ppb. However, the DPR report fails to provide any monitoring data that supports this assumption.

Other exposure scenarios not covered by DPR include elevated exposures that occur when multiple agricultural fields are treated during the same time period in one area (e.g., many strawberry fields treated simultaneously or consecutively in Salinas county). It is possible that workers and individuals living near the treated fields could experience higher exposure levels than predicted by the permit conditions. For example, there are no data on 6-week to 3-month exposures that individuals who live in agricultural areas might receive if multiple methyl bromide applications occur during a season. A reasonable worst-case scenario could be described as multiple nearby fields being treated simultaneously with the air mass moving towards a residential neighborhood located in a lower area of a valley. The subcommittee is not confident that under these conditions, exposures of children and adults to methyl bromide concentrations above the 6-week reference concentrations of 1 and 2 ppb, respectively, do not, or are unlikely to, occur.

Finally, DPR does not address less common exposure scenarios that might occur under unique weather and terrain conditions, such as when a low-level temperature inversion or other similar low-wind condition prevents the dilution of methyl bromide that would normally be expected to occur. Workers and residents living in such an area could be exposed to high methyl bromide concentrations. DPR describes such an exposure scenario in Appendix F, page 253, where 35 bystanders experienced methyl bromide poisoning as a result of low winds and a temperature inversion during and following the applications.

The subcommittee recognizes the difficulty DPR would have in considering all these potential exposure scenarios. However, the subcommittee believes that these likely scenarios need to be evaluated, either by collection of additional monitoring data or by appropriate modeling. Only by doing so can the public have confidence in DPR's assertion that the concentrations to which they are exposed are consistently below regulatory levels.

QUALITY OF DATA AVAILABLE FOR CHARACTERIZING EXPOSURES

This section addresses issues relating to the quality of data available for characterizing exposures, including (1) the analytical methods used in quantifying methyl bromide concentrations; (2) the representativeness of available exposure measurements; (3) the appropriateness of normalizing assumptions used by DPR for different application rates; and (4) the appropriateness of exposure-duration assumptions used in the risk characterization document.

Analytical Methods

The subcommittee has serious concerns about the analytical methods used by DPR and others to determine atmospheric concentrations of methyl bromide. For the most part, these concerns focus on the fact that the initial field-sampling studies were conducted prior to the development of standardized analytical recovery methodologies. In addition, the lack of information on the atmospheric conditions under which the field samples were collected calls into question the recovery values that were used to calculate actual concentrations of methyl bromide in ambient air.

The uncertainty in the recovery values is expressed in a report by Biermann and Barry (1999), which was written after collection of all of the field-sampling data for methyl bromide between 1992 and 1998. Although the analytical method for extracting methyl bromide from the samples had been used previously, it appears that a rigorous testing of the method has not been conducted. The primary uncertainty with the analytical method centers on the procedure for recovering methyl bromide from the charcoal tubes that are used to collect ambient air samples. Prior to the Biermann and Barry (1999) study, recovery values were determined by adding a known amount of methyl bromide in solution to the charcoal, followed by extraction of the charcoal with an organic solvent. It was assumed that addition of methyl bromide in solution to the charcoal was identical to collecting methyl bromide from the gas phase through the charcoal. The percent of methyl bromide recovered from the solution application was considered by DPR to be identical to the percent of methyl bromide recovered from the charcoal in the actual air samples. Biermann and Barry (1999) demonstrated that recovery of air-trapped methyl bromide from the charcoal is only about 50%, whereas the recovery of solution-added methyl bromide from charcoal was reported to be 69% in the field tests. Therefore, the field sample concentrations determined prior to the Biermann and Barry (1999) study were assumed by DPR to have been underestimated by approximately 50%. In its report, DPR calculated the expected concentrations for all the sampling data using the 50% recovery value, based on the Biermann and Barry (1999) study.

The subcommittee considers the 50% recovery estimate of Biermann and Barry (1999) to be questionable for many of the air samples. The 50% recovery estimate is based on samples collected under normal laboratory conditions with ambient air temperatures of between 20°C and 25°C and 20% and 80% relative humidity. However, when Biermann and Barry (1999) took the test system outdoors and did air sampling during the warm daytime temperatures, recoveries were as low as 21% to 26%. In contrast, when the same tests were conducted during the night, recovery estimates were 45% to 48%. Furthermore, when air samples were taken at very low relative humidity (0%), recoveries of methyl bromide were only 0% to 3%. Because relative humidity and air temperature were not considered when the exposure-assessment data were compiled by DPR, and because the sampling data were primarily collected during the daytime, the actual recoveries might be lower than the 50% used by DPR. In addition, the recovery of methyl bromide from the charcoal tubes appears to be dependent on the initial methyl bromide concentration. For instance, in the storage-stability experiment described in Biermann and Barry (1999), recoveries of methyl bromide concentrations at 95 ppb were 5% to 10% lower than recoveries of concentrations at 710 ppb. At even lower concentrations (Biermann and Barry 1999, Table 11), charcoal spiked with 19 ppb methyl bromide yielded 0% recovery; however, only one sample at this low concentration was examined. This 19-ppb concentration was twofold higher than the reporting (detection) limit of the California Department of Food and Agriculture laboratory that did the analysis. The subcommittee is concerned about the lack of reliable recovery estimates at low methyl bromide concentrations, because the reference concentrations (RfCs) for subchronic and chronic exposures to children and adults are 1 and 2 ppb, respectively. The subcommittee believes that DPR and other analytical laboratories might not be able to adequately measure atmospheric concentrations of methyl bromide at or near these RfCs.

The recovery study by Biermann and Barry (1999) provides quantitative information on several environmental factors (e.g., humidity, concentration, temperature) that appear to affect the reliability of ambient air-sampling results of methyl bromide in the field. The field-sampling data presented in the DPR report were collected by at least six different groups, during several time periods (July 1992; October 1992; November 1992; February 1993; and March 1993) and at various locations in California (Santa Maria, Arvin, Chowchilla, Salinas, Hayward, Watsonville, and Madera). Because of the different times and locations at which the air sampling was conducted, it is to be expected that the temperature and humidity levels for each study varied considerably. Daytime temperatures in July and August in the Central Valley of California are often above 100°F, probably near the temperature at which the outdoor recovery study of Biermann and Barry (1999) was conducted, for which reported methyl bromide recoveries were 21 to 26%. Air samples obtained in the cooler months of the year (November-April) were probably collected at temperatures reflective of the 50% recovery of the Biermann and Barry (1999) laboratory samples.

Several of the studies by Siemer and Associates, TriCal, Inc., and AG-Industrial reported that the sampling data was initially adjusted for a recovery of 69% (DPR 1999). However, the DPR report presents no information on whether these 69% recoveries were based on actual samples taken at the time these studies were conducted, or were based on a standardized recovery value. The subcommittee believes that it is unlikely that the 69% recovery used by the several researchers was based on actual laboratory testing, given the uniformity of the recovery estimates. Furthermore, DPR states that, "a field fortification recovery study was not carried out in many of the exposure studies"

(DPR 1999, Appendix H, p. 274). Radian Corporation conducted an additional sampling study and used a slightly different analytical technique (headspace gas chromatography) to determine the methyl bromide air concentrations, but did not report a recovery value (DPR 1999). Air Toxics Limited conducted yet another study using charcoal tubes and a limited number of stainless steel (SUMMA) canisters, which do not have the same recovery problems as charcoal[1]. Recoveries were reported to be in the range of 74% to 125%. Finally, DPR itself conducted residential exposure studies in fumigated houses. Average recoveries were reported to be 71.4%, with a range of 49% to 102%. The location, temperature, and relative humidity for each house appears to be subject to the same variability and uncertainty as for the outdoor air-sampling studies discussed previously.

The analytical data from these studies are clearly compromised by the lack of a robust analytical method for measuring methyl bromide concentrations in air. Because of the ease and lower cost of methyl bromide collection using charcoal as compared with stainless steel canisters, the charcoal method will probably continue to be the method of choice. Therefore, the subcommittee finds that (1) a systematic study should be conducted to assess the usability of the previous sampling data obtained with charcoal tubes and (2) a sampling method should be developed that will provide reliable air concentration data. To accomplish these goals, the following issues should be addressed:

1. Are there types of charcoal (e.g., coconut shell) that give more reliable recoveries than the petroleum-based charcoal used for many of the reported exposure studies?
2. What are the effects of temperature on recovery values? Should the charcoal tubes be maintained at some specific temperature (e.g., 15-18°C) during sampling to minimize degradation of methyl bromide during long (e.g.,12 hr) collection times?
3. Are there methods to minimize the effect of humidity on sample recovery?
4. For each sampling trip, what is the minimum number of samples that should be taken using an alternative method (e.g., stainless steel canisters) to compare recoveries?
5. How does the recovery vary with time of sampling and concentration? What is the limit of detection?

---

[1]Stainless steel canisters are generally evacuated and the sample is captured by allowing air to flow into the canisters. The only surfaces that the methyl bromide comes into contact with in these canisters is the relatively inert stainless steel surface, which is distinctly different from the very large and complex surface of charcoal.

6. A routine method for conducting field-recovery studies should be developed that permits direct air sampling, rather than solvent spiking. Conducting a recovery study using gaseous samples would reduce the uncertainty in the available exposure data.

The subcommittee recognizes the difficulties faced by DPR in using the available sampling data for the exposure assessment. Because the initial field sampling was conducted prior to the critical recovery study, DPR was obliged to use a single recovery study to reevaluate a large number of sampling studies. The air concentrations used in the exposure assessment include an undetermined level of uncertainty due primarily to the uncertainty in the actual analytical recoveries obtained when the samples were collected under field conditions. Nevertheless, the subcommittee feels the data are still very useful and provide important information on methyl bromide emissions from treated areas. With the caveats mentioned previously about air temperature, humidity, and concentration effects on recovery, the 50% adjustment used by DPR appears to be reasonable for most of the samples collected in the cooler months and for concentrations that are greater than 50 ppb. For air samples taken at higher temperatures, the methyl bromide concentrations are probably underestimated, potentially by a factor of 2. If, for example, the outdoor recovery values of 21% to 26% were to prove typical, then the average methyl bromide concentrations would be expected to be about double those estimated by DPR. Because this data set was the primary information used to develop the exposure assessment, and it appears to be the bulk of the information presently available, it is important to place some level of uncertainty on the data. For these purposes, the subcommittee suggests that the actual exposures might be considerably higher than even the adjusted estimates presented by DPR.

Representativeness of Available Exposure Measurements

A representative sample of a diverse group of exposures is a sample that is constructed such that the central tendency (mean) and distribution (standard deviation) of exposure levels observed in the sample are likely to be free of systematic differences from actual exposures that are being assessed. The data presented by DPR reflect a wide variety of occupational exposure scenarios and explicitly represent differences in such factors as soil application methods, depth of application, type of tarping, and soil characteristics. However, even within the occupational exposure groupings, the data indicate very large ranges in exposure concentrations, often of several orders of magnitude. For instance, 24-hr time-weighted average exposures varied widely: for preplant soil injection of methyl bromide they ranged from 0.6 to 835 ppb, for fumigation of grain products from 6 to 6,039 ppb, and for residents downwind of a fumigated house exposures were estimated to range from 40 to 296 ppb. The sources of these variability ranges have not been characterized in the DPR report.

However, aside from these broad ranges in variability, the measurements made for individual scenarios frequently reflect only a single set of samples collected on a single day for one type of exposure. There is little or no discussion in the DPR analysis of how well factors affecting the air sampling, such as air temperature, soil type, wind conditions, and humidity, reflect the actual exposure level distributions in practice for the occupational groups studied. In general, there is an absence of information on the conditions (e.g., temperature, wind conditions, humidity) under which air-concentration measurements were made. Therefore, the subcommittee believes that there is considerable uncertainty about how accurately the observed measurements represent the real distributions of exposure concentrations and durations in the occupational groups that were studied.

Appropriateness of Normalizing Assumptions for Different Application Rates

To estimate occupational exposure levels from soil fumigation, DPR made a simple linear adjustment from the application rates used to the maximum permitted application rates. For example, if the maximum permitted application rate was 400 lb/acre, but only 200 lb/acre was used on the field, where the air concentrations were measured, DPR adjusted the measured air concentrations upward by twofold.

The subcommittee has two reservations about this procedure: the first pertains to the physical transport and transformation of methyl bromide, and the second pertains to the stated goals of the exposure analysis. In the first case, a simple linear adjustment is reasonable if one assumes that the only important mechanisms involved in the transport of methyl bromide between the sites of soil injection and the workers' breathing zones are mixing and dilution, which lead to simple first-order loss independent of concentration. However, if physical sorption to soil particles, and chemical reactions with soil constituents are important, then it is possible that there could be a distribution of sites of high affinity adsorption, or high rate reaction, and that these preferential binding/reaction sites would not be available during methyl bromide soil applications. In this case, methyl bromide applied at higher rates could encounter less effective sorption or reaction in the soil than methyl bromide applied at lower rates, and relatively more methyl bromide could be expected to be available for inhalation by workers. Therefore, there is some risk that the worker exposures at maximally permitted application rates could be somewhat understated.

In the second case, if the goal of the exposure analysis is to represent exposures under the worst-case conditions permitted by the pesticide labels, then the subcommittee agrees that some adjustment for application rates should certainly be made. However, if the goal of the exposure analysis is to represent the distributions of exposure levels that actually exist for the workers, then DPR's goal should be to assure that the exposure data collected appropriately reflect the actual distribution of application rates that are used in practice. If the collected data differ from the exposure distribution being studied, then adjustments should be made to reflect the actual distribution of application rates.

Appropriateness of Exposure-Duration Assumptions

To calculate exposures for durations longer than a single day, DPR has made a large number of assumptions (some of which might be considered relatively conservative) about how many days workers might be exposed at mean levels observed in 1-day studies (DPR 1999, Appendix F, pp. 284-289). The explanation for these assumptions is contained in a single paragraph on page 261:

> Calculations of exposure rely on factors, including application rates, work periods specified in the current California permit conditions, frequency and duration of exposure. Types of tarpaulins, application equipment, and injection depth are used in the permit conditions to determine the maximum daily work time for each type of soil injection fumigation. DPR has requested registrants to provide frequency and duration of exposure for acute and non-acute exposures (Donahue 1997, as cited in DPR 1999). So far, registrants have provided some data as requested. Consequently, default frequency and duration of exposure for many exposure scenarios were generated from data obtained from various sources and the use of professional judgment (Haskell 1998a,b, as cited in DPR 1999). These default values are shown in Appendix A [of the DPR document].

Without more explicit documentation of the specific derivation of the numbers in Appendix F, and the overall goals of this exposure analysis, the subcommittee cannot readily assess the appropriateness of the exposure duration assumptions used.

ACCURACY AND APPROPRIATENESS OF AVAILABLE MODELING TOOLS

Exposure Estimates Based on Modeling

Modeling is an essential tool of risk analysis. It allows us to use our mechanistic understanding of a system to draw inferences about exposure levels and associated risks, even in cases in which we do not have an extensive set of direct observational data. As discussed in more detail below, the subcommittee concludes that in general the basic structure of the residential indoor air dilution and outdoor air dispersion models used in the DPR exposure assessment are appropriate. However, the subcommittee finds that in some cases important questions about the variability of modeled exposures have not been addressed in the DPR report. For example, the subcommittee questions whether DPR has made an appropriate effort to juxtapose model predictions with field observations to characterize the quantitative uncertainties in the model predictions. The subcommittee questions whether DPR has used its models to assess the relevant variability in exposures and risks to different individuals and populations.

DPR presents exposure estimates for individuals in fumigated homes or living near commodity-fumigation facilities in Table 19 (DPR 1999, p.105) of the report. Several of these estimates are based either on regulatory permit levels that are apparently derived from modeling or on model projections themselves; these include exposures of (1) residents in a fumigated house (Table 19-c), (2) residents living near commodity fumigation facilities (Table 19-d), and (3) residents living near fumigated fields (Table 19-e). The modeling approaches supporting each of these cases are addressed below, with (2) and (3) discussed concurrently.

Exposure of Residents in a Fumigated House

The data for the analysis of exposure of residents in a fumigated house were drawn from air concentrations measured in five houses in southern California fumigated on a single day (April 7, 1992) at 1.5 lb/1000 ft^3, and actively aerated using fans for 24 hr before closing the windows. (Data for a sixth house were excluded, reportedly because of a relatively short sampling time.) The data, consisting of a total of 32 methyl bromide concentration measurements made at times ranging from 3 to 92 hr after the end of the initial 24-hr aeration period, are presented in Table 36 (Columns 1 and 2) (DPR 1999, Appendix F). DPR used a single-compartment, simple-dilution model to estimate methyl bromide concentrations after 72 hr of active aeration using the aggregate data from all five homes. This was done by fitting a simple linear regression line to a plot of the logarithm of the observed concentrations versus time. The fitted line (Equation 3-1) is

$$\text{Log(MB)} = -0.008195 \times (t) - 0.148086 \qquad (3\text{-}1)$$
$$r^2 = 0.34955,$$

where MB is concentration of methyl bromide (in ppm), t is number of hours after 24-hour aeration, and r^2 is correlation coefficient.

DPR used this fitted regression line to predict residential exposures for a 1-week period (168 hr) beginning at either 48 or 49 hr after the 24-hr active ventilation (72 hr after the fumigation) without apparent further adjustment for the possibly greater reduction in concentrations that might occur from the 48 additional hours of active ventilation. (DPR requires that active ventilation be carried out for 72 hr after the fumigation, although for these data active ventilation was only done for 24 hr.) To estimate exposure concentrations in northern California, where the fumigation rate is twofold higher (3.0 lb methyl bromide/1,000 ft^3) than in southern California, a simple linear twofold adjustment was made to the methyl bromide concentrations (DPR 1999, Table 36, Columns 4 and 5).

The subcommittee reproduced the regression equation above and derived confidence limits on the rate of exponential decline in methyl bromide concentration over time in Equation 3-2 below.

$$\text{Log (MB)} = -0.008197 \pm 0.00204 \times (t \pm \text{std error}) - 0.1480 \qquad (3\text{-}2)$$
$$r^2 = 0.3497,$$

where MB is concentration of methyl bromide (in ppm), t is number of hours after 24-hr aeration, and r^2 is correlation coefficient.

This regression equation allowed the subcommittee to verify the stated 7-day mean concentrations and associated confidence limits in Table 37 (DPR 1999, Appendix F) of 86 ± 73 ppb (15-229) and 172 ±147 ppb (30-458) for southern and northern California, respectively. It also permitted the subcommittee to determine 24-hr estimates of methyl bromide concentrations to compare with the regulatory target level of 210 ppb that is assumed to apply for the 24 hr immediately following the reentry of residents into their homes. These data are presented in Table 3-1, in which the estimated average methyl bromide concentrations for 1 day and 7 days after the 24-hr ventilation period are shown, along with the standard errors.

A comparison of the subcommittee projections of the central tendency and upper 95% confidence limits for the 7-day average exposure levels (Table 3-1) with the data in Table 37 (DPR 1999, Appendix F) shows that the values correspond closely. However, DPR appears to have made a twofold error in transposing these 7-day results to Table 19c (DPR 1999, p. 105) where the values are given as 172 ± 146 and 344 ± 294 ppb for southern and northern California, respectively.

Aside from the apparent transposition error for the 7-day results, there appear to be deeper problems with DPR's analysis. The grouping of data from five different houses yields, at best, a central estimate of the concentration levels that is likely to be present for residents reentering an average house. This estimate does not reflect the variability among houses in air exchange rates between contaminated wall spaces and the main living areas, and between the living areas and outdoor air. The subcommittee believes that separate analyses of data from each of the five houses would have allowed DPR to make a first-cut assessment of the differences among houses in both initial concentrations of methyl bromide (following 24 hr of active ventilation) and the rates of decline. Because the average methyl bromide concentrations are already relatively high in relation to the regulated target level of 210 ppb (Table 3-1), neglecting this variability raises some concern, although the concern is somewhat mitigated by the fact that DPR apparently made no adjustment for the increased active ventilation period that might occur in practice (i.e., 72 hr versus 24 hr of active ventilation).

Finally, DPR's assumption that the acute 24-hr exposure limit of 210 ppb is achieved is not supported by even the central tendency (median) estimate from the modeled data for the northern California application rate (Table 3-1). This 210-ppb level is based on a calculation that assumes that methyl bromide

TABLE 3-1 Projected 1- and 7-Day Average Methyl Bromide Concentrations (ppb) for Residents Reentering Fumigated Homes

	-2 SE[a]	-1 SE	Median Estimate	+1 SE	+2 SE
Southern California 7-day mean	39	57	87	138	226
Southern California 1-day mean (48 to 72 hr following 24-hr ventilation)	133	175	231	305	404
Northern California 7-day mean	77	114	174	275	452
Northern California 1-day mean (48 to 72 hr following 24-hr ventilation)	266	351	463	611	808

[a]1 and 2 standard error (SE) departures from the central estimate of the regression slope.

concentrations measured at electrical outlets or other enclosed spaces within the wall of a home will be equal to or less than 3 ppm when reentry is permitted. The 210-ppb level also implies that these within-wall measurements accurately reflect the average concentration in a well-mixed wall volume that represents only about 5.6% of the volume of the house, and that the 24-hr average concentrations for the residents reflects immediate mixing of the wall volume contents with those of the living areas of the house, and no loss of methyl bromide from the house during the first 24 hr. Several of these assumptions appear incompatible with the direct observations made from the analysis of the five houses modeled above.

First, the slope of the exponential decline in methyl bromide concentrations (Equation 3-2) reflects a half-life of about 37 hr (with 95% confidence limits of 24 to 73 hr). The average air exchange rate, a general method for expressing ventilation, is 0.019 exchanges per hour (95% confidence limits of 0.009 to 0.028 air changes/hr) in these houses (see Appendix C of this report). This air exchange rate estimate is considerably lower than rates observed in the living areas of other homes. (For example, EPA (1996) reported 24-hr average air exchange rates from approximately 0.33 to 2.2 air changes/hr (10% to 90% range) for 175 houses in Riverside, California.) The low air exchange rates observed for these five homes indicate that the controlling factor for the overall decline of methyl bromide concentrations over time (as observed in DPR 1999, Table 36, Appendix F) cannot be attributed to general house ventilation, but probably reflects slow transfer of methyl bromide between the wall spaces and the living areas. Given this, and the convoluted geometry of wall spaces, the subcommittee questions DPR's assumption that measurements made at one or a few enclosed spaces within the wall are representative of a well mixed space.

The subcommittee also has concerns with the fact that all the data used in the analysis (DPR 1999, Table 36, Appendix F) come from fumigations made on the same day in a similar area of southern California. This means that the data do not account for differences in varying external temperatures, wind conditions, and humidity, and possibly, house structural characteristics in different areas of California and at different times of the year.

Because of the uncertainties surrounding the current data set on exposures of residents returning to fumigated homes, the subcommittee finds that DPR's conclusion that current fumigation practices result in methyl bromide concentrations that do not exceed the regulatory exposure level of 210 ppb does not seem warranted. Further data collection and analysis of exposure concentrations in routinely fumigated homes at different seasons and for different types of homes in various areas of California seems necessary if methyl bromide use as a house fumigant is to be continued with confidence.

Exposure of Residents Downwind from Soil Fumigations

The other major modeling effort in the DPR exposure analysis examines whether residents living near fumigated fields and commodity fumigation facilities are exposed to methyl bromide concentrations that exceed the acute (24-hr) regulatory limit of 210 ppb. These exposures are regulated by an extensive set of permit requirements implemented at the county level and are based on assumptions about the rate of air emissions from soil fumigation operations of various types. A standard air dispersion modeling system, the Industrial Source Complex-Short Term computer model (EPA 1995), is used to calculate the size of buffer zones that are required to prevent methyl bromide concentrations at the boundary from exceeding 210 ppb. DPR used the 210-ppb value to represent the acute exposures of residents near fumigated fields and commodity fumigation facilities in Table 19-d (DPR 1999, p. 105). This 210 ppb value represents an assumption by DPR that the permitting system as currently implemented is working. However, DPR fails to enumerate any underlying conservative assumptions used in their modeling, and does not describe the variability or uncertainty associated with the actual implementation of the permits.

The subcommittee attempted to evaluate DPR's assumption that the 210-ppb exposure level is not being exceeded at the buffer zone boundary. To conduct this analysis, empirical data contained in Table H1 (DPR 1999, Appendix H) of the DPR report were compared with the 210-ppb limit. Table H1 lists 39 maximum methyl bromide concentrations measured between 1992 and 1998 at or near buffer-zone boundaries at field fumigation sites. DPR describes the sampling methodology used to generate these data as follows (DPR 1999, Appendix H, p. 357):

> In these studies, air monitoring was conducted using personal air sampling pumps equipped with activated charcoal tubes. The samplers were set up around the field at a distance of 30 feet from the edge of the field and at the permit condition buffer zone determined for the application. Sampling was initiated at the start of the application and continued for one to seven days, with each sampling interval 6-12 hr. The air flow rate for all samplers was calibrated to approximately 15 mL/min. Wind speed, wind direction, air temperature, and relative humidity were recorded every five minutes with a Met-1 meteorological station.

In Table 3-2, some of the data from Table H1 (DPR 1999, Appendix H) have been reproduced, showing the sampling year, sampling distance, permit condition buffer zone, and methyl bromide concentration (Columns b, c, d,

TABLE 3-2 Maximum Methyl Bromide Air Concentrations from Different Application Methods

(a) Case Number in Table H1	(b) Year	(c) Sampling Distance (ft)	(d) Permit Condition Buffer (ft)	(e) Permit Buffer– Sampling Distance	(f) Sampling Distance as Fraction of Permit Buffer	(g) 24-hr Max MB Concen. (ppm)
1	92	300	390	90	0.77	0.042
2	92	300	330	30	0.91	0.260
3	92	50	330	280	0.15	0.550
4	98	200	200	0	1.00	0.150
5	92	600	1060	460	0.57	0.700
6	92	600	1170	570	0.51	0.610
7	98	510	510	0	1.00	0.110
8	93	200	2010	1810	0.10	0.560
9	93	200	940	740	0.21	0.340
10	95	80	780	700	0.10	0.110
...	...	...	...	...	...	...[a]
30	97	625	420	-205	1.49	0.590
31	92	300	300	0	1.00	0.060
32	96	330	550	220	0.60	1.700
33	97	360	360	0	1.00	0.160
34	97	360	360	0	1.00	0.550
35	98	60	200	140	0.30	0.160
36	98	30	100	70	0.30	0.066
37	98	30	100	70	0.30	0.072
38	98	30	100	70	0.30	0.065
39	98	30	100	70	0.30	0.042
					Mean	0.260
					Std. Deviation	0.332
					Std. Error	0.052
					Geom. Mean	0.145
					Geom. Std. Dev.	2.882

[a]Case numbers 11-29 have been deleted.
Source: Adapted from DPR 1999, Appendix H, Table H-1, pp. 358-360.

and g, respectively). Additional calculations have been made by the subcommittee, including Column e, the absolute distance between sampling distance

(Column c) and the buffer boundary (Column d), and Column f, the ratio of the sampling distance (Column c) to the buffer boundary (Column d). The discussion of these data in the DPR report (DPR 1999) notes that:

> of the 39 applications monitored, seven exceeded the 0.21 ppm target level at the buffer zone distance...Tarpaulin-bedded applications and applications using "very high barrier" tarpaulins appeared to have higher air concentrations than originally assumed in the permit conditions. Of the seven tarpaulin-bedded applications monitored, four exceeded the 0.21 ppm target level at the original buffer zone distance. Of the five very high barrier tarpaulin applications monitored, three exceeded the target level at the original buffer zone distance. None of the other application methods exceeded the target level at the buffer zone distance.

In addition, a footnote in Appendix H (DPR 1999, p. 361) notes that "DPR revised the buffer zones in 1997 and 1998 to provide a higher margin of safety. Under the revised buffer zones, none of the 39 fields monitored exceed 0.21 ppm at the buffer zone distance." Unfortunately, aside from this footnote in the report, no details are provided on the nature and extent of the modifications of buffer zones for the individual cases listed in Table 3-2, nor does DPR indicate how they adjusted the measured data to arrive at their conclusion that none of the cases would have exceeded 0.21 ppm had measurements been taken at the new buffer-zone boundaries.

In Table 3-3, the subcommittee has summarized the data presented in Table 3-2. Methyl bromide concentrations are stratified by distances greater or less than 90% of the buffer zone for pre-1998 and 1998 periods. The data in Table 3-3 suggest that in 1998, methyl bromide concentrations at the prescribed buffer-zone boundaries were lower than those measured prior to 1998. Forty-three percent of the pre-1998 concentrations at the buffer-zone boundary were expected to exceed the regulatory limit of 210 ppb, whereas only 7% of the 1998 concentrations were expected to be over this limit.

Overall, if the 1998 data presented in Tables 3-2 and 3-3 are representative of current permit conditions, the percentage of soil fumigation operations that would result in methyl bromide concentrations at the buffer-zone boundary of greater than 210 ppb is expected to be relatively modest. To make such a conclusion, the subcommittee finds that further data are needed. The 1998 data set of measurements at or near the buffer zones of 30 and 100 feet is very limited. As indicated in Table 3-3, there are only four measurements taken in 1998 at distances greater than or equal to 90% of the buffer zone boundaries. Data collected prior to 1998 suggest that the modeling program estimated methyl bromide concentrations at the buffer-zone boundary that are at or near

TABLE 3-3 Summary Statistics for the Distribution of Methyl Bromide Measurements Near Buffer Zone Boundaries

Sample Group	Pre-1998, distances < 90% of buffer	1998, distances < 90% of buffer	Pre-1998, distances > 90% of buffer	1998, distances > 90% of buffer
Arithmetic mean (ppm)	0.422	0.070	0.286	0.083
Arithmetic standard error (ppm)	0.135	0.012	0.082	0.028
Number of measurements	12	9	13	4
Geometric mean (ppm)	0.239	0.065	0.177	0.067
Geometric standard deviation	3.255	1.469	2.756	2.167
Percent expected to exceed 210 ppb	54	<1	43	7
Projected 90th percentile (ppm)	1.1	0.11	0.65	0.18
Projected 95th percentile (ppm)	1.7	0.12	0.94	0.24
Projected 98th percentile (ppm)	2.7	0.14	1.4	0.33

the 210-ppb limit. This is supported by the arithmetic and geometric concentration means of 0.286 ppm and 0.177 ppm, respectively. However, the subcommittee notes that there is a certain proportion of the measurements that exceed 210 ppb at the buffer-zone boundary, occasionally by several-fold, as indicated by concentrations of 0.65 ppm, 0.94 ppm, and 1.4 ppm at the projected 90th, 95th, and 98th percentiles, respectively.

The subcommittee reviewed two DPR documents that update the material provided in Appendix H of the DPR report (Segawa et al. 2000a,b). These documents provide detailed directions for calculating flux rates and buffer-zone distances for the proposed regulations. Although it is not within the subcommittee's task to comment on the appropriateness of the proposed regulations, it is relevant to the foregoing analysis to note DPR's comparison of buffer-zone distance with monitoring data (Segawa et al. 2000a, p. 8). The authors state that, based on new modeling for 34 applications examined,

> buffer zone table distance was greater than the distance to 0.21 ppm estimated by the ISC [Industrial Source Complex] model for 33 of the 34 applications. On average, the buffer zone table distance exceeded the distance to 0.21 ppm by 520%, with a median of 400% (Table 3). We made these calculations when the monitoring data were originally analyzed using unadjusted air concentrations of the first version of the ISC model. DPR is updating these calculations using adjusted air concentrations and version 3 of the ISC model.

Segawa et al. (2000b) contains a table similar to Table H1 in Appendix H of DPR's report showing maximal concentrations measured at 30 feet, application rates, and proportions calculated to be volatilized using both unadjusted and adjusted measurement recoveries. However, there is no direct presentation analogous to Table H1 of methyl bromide concentrations expected at the revised buffer-zone distances. Therefore, the subcommittee cannot determine the frequency distribution for maximally observed concentrations at the revised buffer-zone distances based on the available information. Accordingly, the subcommittee is unable to fully evaluate the accuracy of the modeling used for estimating off-site residential exposures in the DPR report, nor can the subcommittee determine if the proposed, or even current, buffer zones actually protect nearby residents from exposures to methyl bromide concentrations greater than 210 ppb.

SUMMARY

The DPR report contains a large compilation of exposure data, particularly on worker exposures. However, the subcommittee finds that DPR's exposure analysis is lacking in several respects. Certain exposure scenarios are not dealt with at all in the report, including exposures to residents living near fumigated fields and potentially elevated exposures to residents and workers resulting from methyl bromide application to several fields simultaneously. The subcommittee believes that it is extremely important for DPR to address such exposures, considering that 95% of methyl bromide is used in soil fumigation. Furthermore, there is considerable uncertainty surrounding the analytical recovery methods used in the exposure-assessment studies. Much of the data presented by DPR is based on single air-concentration measurements. There is no discussion of the representativeness of these measurements to the actual exposures experienced by the potentially exposed populations. In addition, DPR makes numerous assumptions regarding durations and levels of exposures, which the subcommittee believes are not explained in sufficient detail to understand their appropriateness. The subcommittee believes that further data collection and analysis are necessary to accurately assess both worker and residential exposures to methyl bromide.

4

Risk Characterization

In this chapter, the National Research Council's (NRC's) subcommittee on methyl bromide considers the material covered in Sections IV and V of the California Department of Pesticide Regulation's (DPR's) risk characterization document. In Section IV, "Risk Assessment," DPR justifies the selection of the toxicological endpoints and the critical no-observed-adverse-effect levels (NOAELs) used in the risk characterization, presents the exposure assessment in the form of tables of exposure measurements for different occupational and residential exposure categories, and presents margins of exposure for each of those categories based on the critical NOAELs and the exposure measurements. In Section V, "Risk Appraisal," DPR addresses the uncertainties in the toxicological and exposure databases, discusses the factors used for intraspecies and interspecies extrapolation, and discusses issues related to the Food Quality Protection Act.

RISK CHARACTERIZATION GOALS

As defined by the NRC (1994) "risk characterization combines the assessments of exposure and response under various exposure conditions to estimate the probability of specific harm to an exposed individual or population. To the extent feasible, this characterization should include the distribution of risk in the population." To properly perform a risk assessment, the hazard posed by the agent must be assessed in terms of the adverse health effects it can cause, the dose-response must be characterized, and the intensity, frequency, and duration of exposure should be determined. The quality of information available for each of these risk characterization components governs the quality of the eventual estimate of risk to individuals by the use of methyl bromide. DPR has addressed each of these risk assessment components in its risk characterization document. In the sections below, the subcommittee reviews DPR's presentation of the information it gathered and analyzed in assessing the risk to agricultural workers and the general population from methyl bromide exposures.

HAZARD IDENTIFICATION

DPR has presented a substantial amount of experimental information on the toxicology of methyl bromide, including the response to various concentrations of the chemical. A number of observations in humans following methyl bromide exposure have been made but preclude a determination of a dose-response relationship. Furthermore, the actual absorbed dosage of methyl bromide is difficult to determine in either animal studies or reports of human exposure. Obviously, in the absence of true dose-response information, the concentration-response is a usable guide for judging, with high confidence, the ambient levels of the chemical that can be expected to represent no harm to humans.

Acute Toxicity Database

The database for the derivation of an acute reference concentration (RfC) includes a single exposure inhalation study with the rat (Driscoll and Hurley 1993), a repeated exposure study with the dog (Newton 1994b), and two well-conducted developmental toxicity studies in different species, the rabbit (Breslin et al. 1990b) and the rat (Sikov et al. 1981). In addition, there is a supporting two-generation reproductive study in the rat and pharmacokinetic studies following inhalation exposure. Therefore, the subcommittee considers the database for the derivation of an acute RfC to be good.

The subcommittee believes that DPR's use of a study with repeated exposures (Newton 1994b) as the critical study on which to base an acute RfC for children is conservative and ensures safety. The NOAEL from a study that uses a single exposure rather than repeat exposures is sufficient to derive an acute RfC provided that there is quantitative dose-response information, the study is conducted with the most sensitive species, and there is a sufficient database of supplemental toxicological information. However, because the rodent study with a single exposure to methyl bromide (Driscoll and Hurley 1993) resulted in a NOAEL that was three times higher than the NOAEL derived from the critical study with the dog (Newton 1994b), the rodent study would have resulted in a less conservative RfC.

With respect to the developmental studies by Breslin et al. (1990a,b), the subcommittee also considered it appropriate for determining an acute NOAEL for the assessment of the risks of acute occupational and residential exposure to methyl bromide. This is plausible given that a single gestational exposure is theoretically sufficient to produce an adverse developmental effect (EPA 1991), particularly blockage of gallbladder development (gestation day 11.5 to 12.5), which occurs over the course of approximately 24 hr in the rabbit. Furthermore, there are large numbers of women of childbearing age in the workforce. Finally, the maternal toxicity that occurred in the Breslin et al. studies (1990 a,b) should not negate the observed developmental effects because gallbladder development occurred 5 to 6 days before the dams displayed toxicity and because only a minority of the dams displayed toxicity. The lowest-observed-adverse-effect level (LOAEL) (80 parts per million (ppm)) used by DPR was based on the dose at which gallbladder agenesis, fetal weight declines, and fused sternebrae were noted. The NOAEL was 40 ppm, resulting in an RfC of 210 ppb.

Subchronic Toxicity Database

In addition to the above-cited developmental endpoints, there were also neurotoxic endpoints selected as critical effects for both the acute and the subchronic time periods. These endpoints were both from a single dog study (Newton, 1994b) in which the dogs at the lowest doses showed signs of depressed activity and the dogs at the higher doses and longer exposure periods showed severe signs of neurotoxicity. Because neurotoxic signs are a prominent feature of human methyl bromide intoxication, this neurotoxicity study in the dog appears to be reasonably selected as the critical study. The acute endpoint was for a NOAEL of 103 ppm, with human equivalent NOAELs of 45 ppm and 25 ppm for adults and children, respectively. Because these were higher than the human equivalent NOAEL calculated from the rabbit developmental study, an RfC was not calculated from this study.

The database for the subchronic studies appears to be quite extensive; there were numerous studies that DPR had an opportunity to evaluate to select a critical study for subchronic toxicity. DPR's selection appears to be appropriate in that they selected a study performed for regulatory purposes that was carefully designed and conducted according to the contract laboratory's standard operating procedures. The one drawback about this study (Newton 1994b) is that it was conducted to establish dose levels for a proposed chronic study (which was subsequently not performed) and not originally planned as a formal subchronic study. As a result, there were decisions made midway through the study by the authors to change the study design with respect to duration or dose levels. There were only four dogs of each sex per treatment group, which is a very small number of replications. The critical observation was made outside the standard operating procedures by a trained veterinarian on two female dogs only, which leaves the observation somewhat equivocal. However, to be conservative, the subcommittee agrees that this still appears to be the most suitable critical study out of a total of 26 studies presented as possibilities by DPR. On the other hand, the subcommittee notes that if these observations on the two dogs at 5 ppm are not considered real manifestations of methyl-bromide-mediated neurotoxicity, selection of another study (e.g., Sikov et al. 1981) as the critical one would raise the NOAEL by approximately an order of magnitude, that is, to 20 ppm (see Table 2-1).

The subchronic estimated LOAEL of 5 ppm is equivocal because of the lack of a dose-response curve at the lower dose levels, the observation of depressed activity in two of eight dogs outside the standard protocol procedures, and the low number of replications. However, the seriousness of the neurotoxicity observed in humans and the potential long-term nature of the neurological effects makes this equivocal observation reasonable as a conservative endpoint.

Chronic Toxicity Database

The existing database identified by DPR for derivation of an RfC for chronic toxicity includes two well-conducted chronic studies with different species, supported by subchronic studies in several species, a two-generation reproduction study, other data on developmental and reproductive effects, and pharmacokinetic studies employing inhalation exposure. The subcommittee considers the database available for derivation of a chronic RfC for methyl bromide good and neither of the key studies had major inadequacies.

The chronic LOAEL (3 ppm) used by DPR was based on the lowest dose that caused changes in the olfactory epithelium in rats exposed to methyl bromide for 29 months (Reuzel et al. 1987, 1991). No effects were observed in the tracheobronchial or pulmonary regions of the respiratory tract and no other exposure-related effects were noted at this concentration. The LOAEL was 30 ppm for all other more adverse effects. The NOAEL and LOAEL for respiratory effects, and also for all other effects, in the NTP study (1992) with mice were 33 ppm and 100 ppm, respectively. The critical endpoint selected from the Reuzel et al. study (1987) is appropriate as more pronounced nasal lesions have been observed at higher concentrations in shorter-term studies (Eustis et al. 1988; Hurtt et al. 1988); however, the critical endpoint in this case is observed at increased incidences only in aged rats, making it an equivocal endpoint. The Hurtt et al. (1988) study indicated that the lesions observed after exposure to methyl bromide at 200 ppm, 6 hr/day for 105 days, were largely reversible. Another consideration is the endpoint of reduced growth of neonatal rats. The NOAEL and LOAEL for reduced growth of neonatal rats in the two-generation reproduction study by American Biogenics Corporation (1986) were 3 and 30 ppm, respectively. Therefore, the Reuzel et al. study (1987) has the lowest LOAEL of the studies considered appropriate for derivation of the RfC. The subcommittee agrees with DPR's choice of this endpoint, with the notation that at the 3-ppm concentration the effects are mild and increased incidences (but not necessarily severity) are observed only in aged rats.

Developmental Neurotoxicity

Methyl bromide is clearly a neurotoxicant in human adults; neurotoxic signs are prominent following high-level human exposures and one study suggests that mild neurotoxic effects might also occur at low levels (Anger et al. 1986). Methyl bromide also is a developmental toxicant as indicated by laboratory animal studies. Therefore, there is reason to suspect that methyl bromide could be a developmental neurotoxicant, which suggests that data from a developmental neurotoxicity test would be informative to the risk assessment. However, the subcommittee finds that the developmental neurotoxicity test, as it is currently described in the U.S. Environmental Protection Agency (EPA) guidelines (EPA 1991), might be inadequate to identify and characterize specific developmental neurotoxicants. Therefore, the utility of data from such a test for DPR's regulatory needs is unclear, and the subcommittee finds that the risk characterization conducted on the currently available database by DPR is probably sufficient for identifying appropriate NOAELs.

EXPOSURE ASSESSMENT

Although DPR has assembled a large data set of occupational exposure studies for methyl bromide, the exposure assessment based on that data set has a number of shortcomings. First, the methyl bromide concentrations in air are compromised by the lack of a robust analytical method for making such measurements. Although the 50% recovery adjustment used by DPR appears to be reasonable for many of the samples, the subcommittee considers it likely that the actual concentrations in air are underestimated rather than overestimated. The measured exposure data for any one occupational exposure category are variable and sparse and nonexistent for residents living near fumigated fields. For approximately one-third of the exposure groups assessed, the data consist of a single measurement. The variability in the exposure measurements reflects the inherent variability in environmental measurements as well as the lack of a comprehensive and systematic sampling program. The subcommittee realizes that DPR was constrained to work with the available monitoring data that was often collected by outside parties, such as growers and manufacturers, for different purposes.

In the exposure assessment, DPR uses various categories of exposure, including acute (24-hr), short-term (7-day), seasonal (subchronic), and chronic. DPR's treatment of these durations and the subcommittee's consideration of them is presented below.

Appropriateness of Acute-Exposure Definition

DPR's use of an acute (24-hr) exposure period is more reasonable for the residential exposure scenario than for an acute occupational (8-hr) exposure scenario. Some individuals, such as infants, young children, or elderly persons, might indeed spend most of a given 24-hr period inside the residence. However, it is unlikely that a worker will be exposed for 24 hr. For the occupational acute-exposure scenario, a shorter duration approximating the length of a work shift (8 hr) would have been more appropriate. This is particularly true for the exposure assessments and the margin of exposure analyses. For example, from the acute neurotoxicity study in dogs (Newton et al. 1994a,b, summarized in Table 3 of DPR 1999, p. 42), it can be seen that a 24-hr exposure to 50 ppm would not be the toxicological equivalent of a 6-hr exposure to 200 ppm. In dogs, a 24-hr exposure to 50 ppm is well tolerated, whereas a 6-hr exposure to 200 ppm would likely cause acute neurological signs. This becomes even more problematic for very short exposures. The 24-hr time-weighted average of a 1-hr exposure to 1,200 ppm is also 50 ppm, but based on the dog neurotoxicity study and the LC_{50} data presented in Table 1 of the DPR report (p. 35), this is likely to be a lethal exposure for at least some of the animals. The subcommittee believes that humans would not respond differently from laboratory animals in this regard.

As a practical matter, because DPR normalized both the methyl bromide concentrations from the exposure assessment studies and the methyl bromide concentrations from the toxicity studies to 24 hr, the end effect might be that the two cancel each other out for the occupational exposure scenario, and the result might lead to a more conservative risk assessment for the residential exposure scenario. This is because the studies that DPR used to determine the critical NOAELs for acute toxicity all used exposure durations of 6 to 8 hr. DPR then normalized the NOAEL concentrations to 24 hr using concentration/exposure duration relationships. Thus, the actual exposure durations used in the studies were good approximations for an acute occupational exposure, and the two normalizations essentially canceled one another out when the margins of exposure (MOEs) were calculated. On the other hand, as already mentioned, a 6-hr exposure to 200 ppm is likely to be more acutely toxic than a 24-hr exposure to 50 ppm. Thus, the 24-hr normalized NOAEL might be lower than a NOAEL derived from actual 24-hr exposures would be. Because the residential exposure measurements should be made over 24 hr, and therefore would not have to be normalized, the MOEs for the residential exposure scenarios would be more conservative than the occupational MOEs. As discussed in Chapter 3 of this report, the fact that few actual exposure measurements were made for the residential exposure scenarios is a separate problem with the risk assessment.

Subchronic Exposure

DPR defines two categories of subchronic exposure: short-term and seasonal. The subcommittee agrees with DPR that it is appropriate to have a subchronic exposure category to describe worker exposures in preplant soil fumigation and commodity fumigation, and that a subchronic category might also be appropriate for residents of fumigated houses or residents who live near fumigation facilities. As outlined in Section IV of the DPR report (Table 19, p. 105) residents might have short-term exposures by virtue of moving back into a fumigated house. Residents may also have seasonal exposures as a result of living near fumigation facilities. The subcommittee also believes that it is plausible that residents living near fumigated fields might be exposed to methyl bromide for periods lasting longer than 24 hr, and therefore, that Section IV should include an exposure assessment for short-term and seasonal exposures to residents near fumigated fields.

The subcommittee does not believe that it is appropriate to assume, based on the short half-life of unmetabolized methyl bromide, that the effects of methyl bromide are completely reversed a few days after cessation of exposure. The subcommittee bases this statement on the fact that toxicology studies suggest that longer exposures are associated with lower NOAELs than shorter studies, indicating that some processes involved in methyl bromide toxicity are not likely to be quickly (within a few days) or completely reversible. Therefore, the subcommittee concurs with DPR that the seasonal exposure category is an appropriate one for workers who have repeated exposures to methyl bromide, separated by periods up to several days, over the course of a season.

As stated in Table 15 (DPR 1999, p. 92) and the description of the exposure durations in the DPR report (DPR 1999, p. 10), the durations for the short-term and seasonal scenarios appear to be 1 week and 6 weeks, respectively. These are the treatment durations at which effects were observed in the two critical subchronic studies identified by DPR (Sikov et al. 1981; Newton et al. 1994b). However, elsewhere in the document, DPR states that the seasonal exposure duration is "greater than one month" (DPR 1999, p. 90), and still elsewhere as 90 days (DPR 1999, Section IV.B, p. 93). The subcommittee believes that the appropriate duration is 90 days because the seasonal uses noted above are likely to last longer than 1 month or 6 weeks. The distinction between the duration of the critical studies and the actual durations of exposure for the scenarios should be clarified in Tables 15 to 20. The subcommittee concurs with the point made in the DPR report (p. 90) that the seasonal NOAEL might have been lower if the dogs in the critical study had been exposed for longer than 6 weeks (Newton et al. 1994b). However, as already discussed in this report, the subchronic RfC derived from that study is a fairly conservative one, and therefore, probably protective even for longer exposure durations.

Chronic Exposure

Chronic exposure generally refers to a 70-year (lifetime) continuous exposure to the chemical of concern. There does not appear to be chronic nonoccupational exposures for any populations associated with field agricultural applications, because the application of methyl bromide is on a seasonal basis, not year round. However, the subcommittee believes that chronic nonoccupational exposures could be possible for residents near commodity-fumigation facilities or transport facilities, where fumigation of commodity storage warehouses or shipping containers might occur on a year-round basis. Fumigation workers also might have chronic exposures.

The subcommittee believes that DPR's normalization of the 6 hr/day, 5 days/wk, exposure of the lifetime study for rats to a 24-hr/day lifetime exposure for humans is appropriate, but notes that it adds another layer of conservativeness to the derived value. The RfC should be applied to lifetime exposures. The subcommittee disagrees with DPR's definition of chronic exposure for humans as "a year or more" (DPR 1999, p. 4), because this definition does not agree with the accepted EPA definition (EPA 1989) of chronic exposure as a period between 7 years (approximately 10% of a human lifetime) and a lifetime. Subchronic exposures are defined by EPA as ranging from several months to several years.

MARGIN-OF-EXPOSURE ANALYSIS

DPR has done a tremendous amount of work in pulling together a very large amount of exposure information to compare with methyl bromide concentrations and durations of toxicological concern. DPR has chosen to use an MOE, a ratio of the critical human equivalent NOAEL to the estimated human exposure levels, to characterize the risks posed to agricultural workers, nearby residents, and residents returning to fumigated homes. Nevertheless, the MOE analysis is one of the least satisfying aspects of the DPR document.

The risk characterization document, as reviewed by the subcommittee, contains neither a statement of DPR's information objectives or data needs, nor does it indicate how the MOE methodology used is related to those needs. There is minimal quantitative treatment of variability and no apparent quantitative analysis of uncertainty (both discussed below). The subcommittee believes that it is critical that DPR explicitly state how these important issues could affect the analysis to produce information that is helpful for decision-making. DPR appears to be using the exposure data to make a large number of binary comparisons (e.g., safe and dangerous) directly from the observed data, with adjustments to the maximum permissible application rate, and assumptions about the repetition of exposures from day to day. The level of concern for safe or dangerous exposure is an MOE of 100; when MOEs are greater than 100, the populations are assumed to have little risk of adverse effects, and when the MOEs are less than 100, there is a cause for concern for potential adverse effects. DPR appears to be asking, "Do the single-day exposure data that have been directly observed for particular groups, such as applicators, indicate that when methyl bromide is used at the maximum permissible application rate, these workers or residents will be exposed to concentrations that provide a less than a 100-fold margin below the projected human-equivalent NOAEL for acute exposures (21 to 45 ppm)?" and "Do the acute exposure data indicate a less than 100-fold margin for the longer-term endpoints based on DPR's assumptions for weekly and seasonal exposures?" In all, the Tables 16-19 (pp. 96-105) of the DPR report give exposure data for

160 different worker and residential groups, and an additional five cases are based on modeling. Even within a specific exposed group, exposure levels for a particular duration are both considerably variable and, depending on the database, uncertain.

"Variability" in modern risk assessment is defined as real differences among cases (Cullen and Frey 1999; Hattis and Anderson 1999; Hattis and Barlow 1996; Hattis and Burmaster 1994; Thompson 1999). Breaking the data down into different kinds of exposed groups, as in DPR's set of 160 exposure categories, addresses one source of variability. However, characterizing the real variability in exposures experienced by different people within an exposed group is also critical for an informative risk evaluation. This especially applies for a toxicant with a highly upward turning nonlinearity in its population dose-response curve because the individuals at the high-end concentration of the exposure distribution are generally at much larger potential risk for more serious adverse effects than more typical members of the exposed group. Variability in exposures is usually characterized by some measure of dispersion, such as the geometric standard deviation for usual unimodal lognormal distributions.

"Uncertainty," in contrast, reflects the imperfection in our knowledge about the true value of a parameter—including parameters characterizing variability. Uncertainty can be reduced by better and more extensive data, improved models, and so forth. Characterizing uncertainty is important in a risk analysis to frankly convey how confident the audience should be in the results and conclusions presented. Commonly used measures of uncertainty include the standard error of the mean or the standard error of the estimate of a regression coefficient in a standard multiple regression analysis (Cullen and Frey 1999; Hattis and Anderson 1999; Hattis and Barlow 1996; Hattis and Burmaster 1994; Thompson 1999).

Of the 160 worker categories presented, the exposure estimates for 59 categories are based on a single air-concentration measurement; 43 categories are assessed based on only two measured air values, and the remaining 58 categories had more than two measurements. The subcommittee finds that the treatment of data from worker and general population groups with these differing amounts of data is neither consistent nor designed to produce useful estimates of exposures of concern with respect to variability and uncertainty. The subcommittee comments on each of these cases below.

Categories with More than Two Data Points

DPR has summarized acute exposures for worker categories when there are more than two data points as the range of the data directly observed (after adjustments for such things as application rates) plus a simple arithmetic mean and arithmetic standard deviation. However, DPR did not appear to calculate a consistent percentile from the observed data. In the current analysis, DPR has implicitly treated with greater conservatism cases in which there are more data points (a higher percentile is used) than cases in which there are fewer data points. Moreover, basing calculations on the highest of N values introduces statistical instabilities into the analysis. Finally, DPR has not provided a rationale for their choice of arithmetic means and arithmetic standard deviations, rather than more typical lognormal statistics. Analyses of data for some groups (e.g., copilots, applicators, tarp removers) by the subcommittee (see Appendix C) indicates that in general lognormal distributions would be more appropriate than would normal distributions, as is usual for exposure distributions. (For further discussion of the use of lognormal distributions for describing the variability in exposures, see Cullen and Frey (1999); Hattis and Burmaster (1994); Thompson (1999).)

Categories with One or Two Data Points

In most cases, DPR has calculated a mean (if there were two points) and listed the higher of two points as the high value. However, in some other cases, a "95th percentile" value is calculated by assuming an arithmetic standard deviation equal to the mean of the one observed data point. DPR does not explain why this is done for some single-point exposure categories and not others. For cases in which there are only one or two data points, the subcommittee encourages DPR to either gather additional data or consolidate related exposure categories on an a priori basis (i.e., not based on the measured levels but based on similarity of the processes generating the exposures) to assemble greater numbers of data points for basic statistical treatment within groups.

UNCERTAINTY ISSUES

In Section V, "Risk Appraisal," of the DPR report, DPR discusses the limitations of its risk characterization for methyl bromide and how it quantitatively and qualitatively dealt with the specific uncertainties The subcommittee comments upon these limitations and DPR's approach to them below.

Derivation of Reference Concentrations

DPR has developed inhalation RfCs for acute, subchronic, and chronic exposures. When derived from NOAELs, the RfCs reflect 100-fold uncertainties, with a 10-fold uncertainty for species differences and a 10-fold uncertainty for variations among humans. When NOAELs were estimated from LOAELs, a 10-fold uncertainty was used. DPR states that it is their policy to use a default 10-fold uncertainty factor to estimate a NOAEL from a LOAEL. The subcommittee in general agrees with this application of a default uncertainty factor of 10, particularly where the endpoint is an adverse effect such as neurotoxicity or developmental toxicity. In the case of the chronic RfC, the LOAEL was so mild as to be close to a NOAEL. In that case, the subcommittee suggests that a three-fold uncertainty factor be considered. However, the subcommittee realizes that DPR is constrained to have a chronic RfC no higher that the EPA's chronic RfC of 1.3 ppb. Therefore, DPR's chronic RfC for adults of 2 ppb is reasonable. The subcommittee also agrees that the interspecies and intraspecies uncertainty factors of 10 (for each) applied to the acute and subchronic RfCs are appropriate.

The subcommittee notes that DPR used an older method than EPA's current method for the derivation of its RfCs (EPA 1994). DPR derived separate values for adults and children; this is not possible with the current EPA method. It is interesting to note that although different methodologies were used, the RfCs derived by DPR for adults and children, 2 ppb and 1 ppb, respectively, are similar to EPA's value of 1.3 ppb.

Sensitive Subpopulations

The Food Quality Protection Act of 1996 mandated that the EPA use an additional 10-fold safety factor for infants and children, unless it could be determined from available data that a different factor would be safe. The subcommittee considered the methyl bromide database in light of the three criteria used by EPA to determine the safety factor. The first criterion concerns the completeness and reliability of the toxicology database. As discussed at length in Chapter 2 and earlier in this chapter, the subcommittee finds that the toxicology database for methyl bromide was good overall. The second criterion concerns the completeness and reliability of the exposure database. As discussed in Chapter 3, the exposure database, though flawed, is quite extensive for occupational exposures. In contrast, for residential exposures, the category into which most exposures to children and infants would fall, the database is inadequate. Limited data are available only for residential exposures during fumigation of the residence, not for residents living next to fumigated fields or fumigation facilities. The final criterion concerns the potential for prenatal and postnatal toxicity. The two-generation rat reproduction study (American Biogenics Corporation 1986) and the rabbit and rat developmental toxicity studies (Breslin et al. 1990a,b; Sikov et al. 1981) indicate that methyl bromide is not a potent teratogen, but that it can cause developmental toxicity. The teratogenic effect, gallbladder agenesis, is considered to be a minor malformation and this effect was seen only at doses that caused maternal toxicity. The subcommittee expects that a potent teratogen would cause multiple malformations at doses that do not cause maternal toxicity. As noted in the DPR report, there is some evidence for increased sensitivity of the developing organism to adverse effects of methyl bromide compared with the mothers in rats (American Biogenics Corporation 1986), but not in rabbits (Breslin et al. 1990a,b). Although good otherwise, the reproductive and developmental database lacks a developmental neurotoxicity study. According to DPR (DPR 1999, p. 126), EPA has added an additional uncertainty factor of 3 in the absence of such a study in its recent time-limited tolerances for pesticides.

Given that the NOAELs used for the various exposure scenarios are already quite conservative, the subcommittee felt that an additional safety factor for infants and children was not necessary.

Multiple Exposures

Although DPR acknowledges that workers might receive multiple exposures from methyl bromide, there is only a limited discussion on the potential exposure of residents who live in areas where multiple fields might be fumigated simultaneously or within a short period of time. Because the majority of methyl bromide is used in field applications, residents near treated fields are subject to frequent exposures during the fumigation season. The subcommittee notes that it would be unrealistic to assume that most residents in agricultural areas live near only one treated field. Therefore, the buffer zones established by DPR to be protective of residents adjacent to one field might not be sufficient should the residents be near multiple treated fields. Although these exposures were commented upon in Chapter 3, the subcommittee reiterates that this is a significant data gap in the exposure assessment.

In addition, the subcommittee has concerns regarding repeated exposures of workers, such as soil or structural fumigators, because soil fumigators might have repeated exposures on consecutive days for several months or structural fumigators might be engaged in multiple fumigations on a single day (Anger et al. 1986). The potential that such repeated exposures might occur raises concerns in light of results from Anger et al. (1986) that suggest that relatively low exposure levels (<2 to 3 ppm) of methyl bromide from fumigation might produce slight neurotoxic effects in workers. Additional data on the neurotoxic effects of methyl bromide in exposed workers are needed.

The proposed regulations provide for a 36-hr waiting period between the application of methyl bromide to a field near a school and when school is in session. In practical terms, this means that fumigation of fields near schools are limited to Friday evenings and Saturday. However, DPR makes no provision for school activities that might occur during weekends at the school, particularly outdoor activities such as sports. Such exposures should not be ignored, because children might have greater susceptibility to effects from methyl bromide exposures, and because data suggest that slight neurotoxic effects might occur at low concentrations (Anger et al. 1986).

SUMMARY

DPR characterized the risks associated with exposure to methyl bromide by using an MOE approach. The subcommittee found this approach to be reasonable for determining which workers or residents are likely to be exposed to potentially harmful methyl bromide concentrations. However the subcommittee had concerns about DPR's use of MOEs for risk characterizations and for protecting nonworkers, in particular, people living near fumigated fields. DPR has not indicated how the MOEs are to be used to determine the protectiveness of the buffer zones specified in the application permits. Nor has DPR characterized certain potentially sensitive populations, such as children in schools or living near fumigated fields, although the proposed regulations address the exposure of children by restricting the application times near schools. The subcommittee feels that the uncertainties addressed by DPR in the report, including extrapolating from LOAELs to NOAELs, and from animals to human, although important, are only part of the uncertainties that need to be dealt with in the document.

5

Conclusions and Recommendations

The California Department of Pesticide Regulation (DPR) has put considerable time and effort into the development of its risk characterization document for methyl bromide. The subcommittee agrees that development of a risk characterization, and subsequent risk assessment, is an appropriate approach to be used to protect agricultural workers and the general population from potential adverse effects associated with this widely used pesticide. Below are specific conclusions reached by the subcommittee based on DPR's presentation of the toxicology, exposure, and risk assessment and risk appraisal information for methyl bromide as detailed in DPR's report. Recommendations on improving both the data quality and the analytical approaches used in the risk assessment are presented as a means to assist DPR in identifying at-risk populations and, subsequently, developing regulations to protect them.

TOXICOLOGICAL INFORMATION

Conclusions

- The subcommittee agrees with DPR's selection of the toxicological endpoints and the NOAELs used to derive the inhalation reference concentrations (RfC). The subcommittee considers the NOAELs to be protective and conservative.

- The subcommittee agrees that it is appropriate to use a developmental study for the derivation of an acute RfC for the general population.
- DPR's selection of the dog study (Newton 1994b) with a neurotoxicity endpoint is appropriate for developing a subchronic RfC, but the subcommittee is concerned about whether the decrease in responsiveness seen at exposure to 5 ppm of methyl bromide in two of eight dogs is a true LOAEL or even an effect at all. Nevertheless, the subtle neurological deficits observed in occupationally exposed humans (Anger et al., 1986) supports these animal data that neurotoxic responses can occur at low exposure concentrations. Therefore, the subcommittee concurs with the conservative assignment of the 5 ppm value in this dog study as a LOAEL.
- The rabbit developmental study had toxicity endpoints of gallbladder agenesis and fused sternebrae, which are not considered major malformations; however, the subcommittee feels that these are indicators of developmental toxicity, and therefore, are appropriate endpoints for the developmental RfC (Breslin et al. 1990b).
- The subcommittee agrees with DPR's selection of nasal epithelial hyperplasia as the toxicity endpoint for the chronic RfC, but notes that the effect is mild and might be closer to a NOAEL than a LOAEL.
- In general, DPR's presentation of the toxicological information is clear and easy to follow and permits the reader to follow DPR's logic in selecting critical studies and NOAELs.

Recommendations

- Methyl bromide is a methylating agent that is a direct-acting mutagen in vitro. However, there are good animal studies that indicate it is not carcinogenic. DPR should review the literature for any discussion on methyl bromide and other methylating agents as to why an in vitro mutagen is not an in vivo carcinogen. This could aid in understanding the mechanism of methyl bromide toxicity and lend confidence when extrapolating from the animal data to humans.
- The dog study from which the 6-week subchronic RfC is derived (Newton 1994b) had a small number of animals and some subjective observations that led to a LOAEL of 5 ppm. The subcommittee recommends that a new study be conducted to verify the neurotoxicity endpoints of decreased responsiveness at 5 ppm.
- Further developmental studies on methyl bromide would help to clarify several major issues
    — Does in utero or early postnatal exposure to methyl bromide affect

adult reproductive function? This question arises from the observation of apparently reduced fertility in the F1 offspring, but not the F0 parents in a two-generation study (American Biogenics Corporation 1986; Hardisty 1992; Busey 1993).

— What are the critical exposure periods for expression of reduced pup weights found during lactation and decreased offspring brain weights and dimensions (i.e., are they due to gestational or lactational exposure to methyl bromide?)

— Is methyl bromide excreted in breast milk? This question could be answered by measuring methyl bromide concentrations in the breast milk of lactating animals exposed to methyl bromide by inhalation.

— Does gallbladder agenesis occur following a single exposure to methyl bromide during the critical period for gallbladder development?

EXPOSURE ASSESSMENT

Conclusions

- Although the exact levels of exposure for workers and residents are not known, DPR has collected a substantial amount of data that indicate that some of these exposures are significant, exceeding current regulatory limits, and therefore are of concern.

- The measures of exposure are frequently based on a single value with no accompanying information on ambient air temperature, relative humidity, and wind conditions. The lack of representativeness of the measurements is not assessed in the main text of the DPR report and is only acknowledged as a possible confounder in an appendix.

- In general, the subcommittee is highly critical of the analysis and presentation of the available exposure data, finding it seriously deficient in understanding and application of modern concepts of variability and uncertainty, and in the fair evaluation of the magnitude and distribution of existing exposures relative to exposure levels intended to be achieved by current regulatory controls.

- There is considerable room for improvement in the methods used by DPR to obtain monitoring data, particularly with regard to good measurement techniques and sampling strategies that assess variability of actual exposure.

- Information is lacking on exposures to residents living near application areas and exposures for populations subject to aggregate applications (e.g., those living in basin area where multiple fields have been treated). Available data and modeling suggest that for some populations, exposures might exceed regulatory limits.

- A substantial ambiguity exists for current methods used to measure methyl bromide in air, particularly with respect to recovery values and the field conditions during air sampling. As a result, actual measured air concentrations of methyl bromide and potential exposure levels are uncertain.
- DPR's use of 24-hr averaging for 8-hr exposures adds a further uncertainty to the protectiveness of the regulations.
- DPR's documentation of their exposure assessment is difficult to follow and requires searching through numerous appendices and other documents (many of which were requested by the subcommittee at a later date) to determine the data sources used by DPR and the approach that was used to evaluate and model the data. A roadmap of the information in the appendices and a more systematic presentation of the data would be helpful to the reader. In particular, DPR's discussion of buffer zones and the measurements taken at them, is confusing and appears to be missing important pieces of information.

Recommendations

- DPR should explicitly state what populations or subpopulations are expected to be represented by the scenarios.
- Identify the best analytical methods for determining methyl bromide concentrations in air under a variety of field conditions. The entire risk assessment process is fundamentally dependent on the quality of the analytical information on exposure conditions. A substantial effort is needed to develop rigorous and robust field analytical methods for determining concentrations of methyl bromide. This will require a complete multilaboratory series of tests that can allow a determination of the reliability of analytical information from field samples.
- Conduct systematic recovery analyses of field and laboratory air samples under a variety of air temperature, wind, and relative humidity conditions.
- Establish a new sampling program to determine the representativeness of exposure data with an emphasis on residential (including house fumigations) and high-exposure occupations.
- DPR should consider quantifying—at the very least—the potentially exposed populations in its occupational categories, and if possible, the number of residents near fields, fumigation facilities, and residents returning to fumigated homes.
- DPR should evaluate its exposure data using modern distributional concepts—including both variability and uncertainty to quantify how accurately the observed measurements represent the real distributions of exposure concentrations and durations. The subcommittee believes that analyses intended to support regulations should frankly disclose the expected degree of confi-

dence the public should have that real exposures will be kept below regulatory levels for defined percentiles of exposed populations.

RISK CHARACTERIZATION

Conclusions

- The subcommittee overall agrees with the risk characterization for inhalation exposure of methyl bromide. The subcommittee believes that the toxicity endpoints used might be overly conservative due to their equivocal nature, but also believes that the exposure assessments might understate the actual exposures, particularly for residents living near fields where methyl bromide is applied.
- The subcommittee agrees that DPR's use of factors of 10 for intraspecies variation and for animal to human variation, as well as a benchmark margin of exposure (MOE) of 100, is consistent with traditional risk management practices.
- The subcommittee believes that the uncertainties associated with DPR's exposure levels call into question the validity of its MOE values. Given the likelihood that the error in the measurements will underestimate some exposures, the subcommittee anticipates that some MOEs will be lower than those calculated by DPR, some of which already indicate there is a cause for concern (i.e., they are currently less than 100).
- Given the lack of information on methyl bromide drift off-site from fumigated fields, it is unclear to the subcommittee how DPR can develop a coherent and protective plan for buffer zones and injection times for field fumigation as specified in Section 6450 of Title 3 of the California Code of Regulations.
- The subcommittee concludes that DPR has failed to conduct a true risk assessment in that it does not combine both exposure assessments and dose-response assessments to estimate the probability of specific harm to exposed individuals or populations. Furthermore, DPR does not characterize the distribution of risk to the exposed populations.

Recommendations

- Buffer zones should be derived so that they appropriately protect those persons who might spend appreciable amounts of time near treated areas (e.g., residential, schools, offices). These buffer zone distances will need to be based on reasonable worst-case scenarios. Additional field studies should be undertaken to validate these buffer zones.

- At the very least, DPR should characterize occupational and residential exposures with distributions, that is, estimate how many people are likely to be exposed at what levels relative to levels of concern for a given duration of exposure. DPR should also conduct some uncertainty analyses to determine what level of confidence in the exposure values is appropriate given the existing data.

- More neurological testing among those occupationally exposed, particularly at various time intervals after methyl bromide exposures have occurred (instead of during exposures), would enable DPR to look for possible long-term or permanent effects.

- To protect workers and residents from the adverse effects of methyl bromide, DPR must be more explicit about linking its methodology for exposure and MOE analysis to the regulatory levels that are based upon the risk assessment or MOE values. The subcommittee recommends that DPR state at the beginning of its risk characterization document the regulatory goals it hopes to achieve and how its risk characterization will meet them.

In conclusion, the subcommittee recognizes that conducting additional toxicity testing and exposure monitoring is somewhat problematic given the phase-out of methyl bromide in the United States by 2005. Nevertheless, the subcommittee believes that extensive use of this pesticide at this time in California and elsewhere warrants an acknowledgment of existing data gaps that must be addressed to ensure that agricultural workers and residents living near areas where methyl bromide is used are protected against the short-term and long-term health effects of this pesticide. These data gaps might require the combined efforts of regulatory agencies such as DPR and the methyl bromide industry, including manufacturers and pesticide applicators.

References

ACGIH (American Conference of Governmental Industrial Hygienists). 1997. Methyl Bromide in: Supplements to the Sixth Edition Documentation of the Threshold Limit Values and Biological Exposure Indices. Cincinnati, OH: ACGIH.

American Biogenics Corporation. 1986. Two-Generation Reproduction Study Via Inhalation in Albino Rats Using Methyl Bromide. Study 450-1525. Submitted to Methyl Bromide Panel, c/o Great Lakes Chemical Corporation, W. Lafayette, IN. American Biogenics Corporation, Decatur, IL.

Anger, K.W., L. Moody, J. Burg, W.S. Brightwell, B.J. Taylor, J.M. Russo, N. Dickerson, J.V. Setzer, B.L. Johnson and K. Hicks. 1986. Neurobehavioral evaluation of soil and structural fumigators using methyl bromide and sulfuryl fluoride. Neurotoxicology 7(3):137-156.

Bentley, K.S. 1994. Detection of Single Strand Breaks in Rat Testicular DNA by Alkaline Elution Following in Vivo Inhalation Exposure to Methyl Bromide. Haskell Laboratory Report No. 54-94. Chemical Manufacturers Association. DPR Vol. 123-188 #162362.

Biermann, H.W. and T. Barry. 1999. Evaluation of Charcoal Tube and SUMMA Canister Recoveries for Methyl Bromide Air Sampling. Environmental and Pest Management Branch. DPR. EH99-02 (Appendix K of DPR 1999).

Bond, J.A., J.S. Dutcher, M.A. Medinsky, R.F. Henderson, and L.S. Birnbaum. 1985. Disposition of ^{14}C methyl bromide in rats after inhalation. Toxicol. Appl. Pharmacol. 78(2):259-267.

Bonnefoi, M.S., C.J. Davenport, and K.T. Morgan. 1991. Metabolism and toxicity of methyl iodide in primary dissociated neural cell cultures. Neurotoxicology. 12(1):33-46.

Breslin, W.J., C.L. Zablotny, G.J. Bradley, K.D. Nitschke, and L.G. Lomax. 1990a. Methyl Bromide Inhalation Teratology Probe Study in New Zealand White Rabbits. Final Report. K-000681-032. Study prepared by The Toxicology Research Laboratory, The Dow Chemical Company, Midland, MI for The Methyl Bromide Industry Panel, Chemical Manufacturers Association, Washington, DC.

Breslin, W.J., C.L. Zablotny, G.J. Bradley, and L.G. Lomax. 1990b. Methyl Bromide Inhalation Teratology Study in New Zealand White Rabbits. Final Report. K-000681-033. Study prepared by The Toxicology Research Laboratory, The Dow Chemical Company, Midland, MI for The Methyl Bromide Industry Panel, Chemical Manufacturers Association, Washington, DC.

Busey, W.M. 1993. Neuropathological Evaluation of Brains from F0 and F1 Rats in Two-generation Reproduction Study with Methyl Bromide Pathology Report. EPL project number 303-007. DPR Vol. 123-153 #125516.

Chahoud, I., A. Ligensa, L. Dietzel and A.S. Faqi. 1999. Correlation between maternal toxicity and embryo/fetal effects. Reprod. Toxicol. 13(5):375-381.

Cullen, A.C. and H.C. Frey. 1999. Probabilistic Techniques in Exposure Assessment. A Handbook for Dealing with Variability and Uncertainty in Models and Inputs. New York: Plenum Press.

Disse, M., F. Joo, H. Schulz and J.R. Wolff. 1996. Prenatal exposure to sodium bromide affects the postnatal growth and brain development. J. Hirnorsch. 37(10):127-134.

Djalali-Behzad, G., S. Hussain, S. Osterman-Golkar, and D. Segerback. 1981. Estimation of genetic risks of alkylating agents. VI. Exposure of mice and bacteria to methyl bromide. Mutat. Res. 84(1):1-9.

Donahue, J.M. 1997. DPR letters dated November 12. (Letters were sent to six MB registrants requesting data on frequency and duration of exposure). WH&S. DPR.

DPR (Department of Pesticide Regulation). 1992. Toxicology Review Prompts Changes in Structural Fumigations. Release no. 92-07. Department of Pesticide Regulation, California Environmental Protection Agency.

DPR (Department of Pesticide Regulation). 1999. Methyl Bromide, Risk Characterization Document for Inhalation Exposure. Draft RCD 99-02. Department of Pesticide Regulation, California Environmental Protection Agency. October 1999.

Driscoll, C.D. and J.M. Hurley. 1993. Methyl Bromide: Single Exposure Vapor Inhalation Neurotoxicity Study in Rats. Lab. Project ID 92N1197. Study prepared by Bushy Run Research Center, Union Carbide Chemicals and Plastics Company, Export, PA for The Methyl Bromide Industry Panel, Chemical Manufacturers Association, Washington, DC.

EPA (U.S. Environmental Protection Agency). 1989. Risk Assessment Guidance for Superfund, Vol. I. Human Health Evaluation Manual, Part A. EPA/540/1-89/002. Washington, DC: EPA, Office of Emergency and Remedial Response.

EPA (U.S. Environmental Protection Agency). 1991. Guidelines for Developmental Toxicity Risk Assessment. EPA/600/FR-91/001. Washington, DC: EPA, Office of Research and Development.

EPA (U.S. Environmental Protection Agency). 1994. Methods for Derivation of Inhalation Reference Concentrations and Application of Inhalation Dosimetry. EPA/600/8-90/066F. Washington, D.C.: EPA, Office of Research and Development.

EPA (U.S. Environmental Protection Agency). 1995. User's Guide for the Industrial Source Complex (ISC3) Dispersion Models. Volume 1. User Instructions. U.S. EPA Office of Air Quality Planning and Standards; Emissions, Monitoring and Analysis Division, Research Triangle Park, North Carolina. Online. Available: www.epa.gov/ttn/scram/

EPA (U.S. Environmental Protection Agency). 1996. The Particle TEAM (PTEAM) Study: Analysis of the Data: Final Report. Volume III. EPA/600/R-95/098. Research Triangle Park, NC: EPA, Office of Research and Development.

EPA (U.S. Environmental Protection Agency). 1997. Good Laboratory Practices. 40 Code of Federal Regulations, Part 160.

Eustis, S.L., S.B. Haber, R.T. Drew, and R.S. Yang. 1988. Toxicology and pathology of methyl bromide in F344 rats and B6C3F1 mice following repeated inhalation exposure. Fundam. Appl. Toxicol. 11(4):594-610.

Gansewendt, B., U. Foest, D. Xu, E. Hallier, H.M. Bolt, and H. Peter. 1991. Formation of DNA adducts in F-344 rats after oral administration or inhalation of [^{14}C] methyl bromide. Food Chem. Toxicol. 29(8):557-563.

Garnier, R., M.O. Rambourg-Schepens, A. Muller and E. Hallier. 1996. Glutathione transferase activity and formation of macromolecular adducts in two cases of acute methyl bromide poisoning. Occup. Environ. Med. 53(3):211-215.

Gibbons, D.B., H.R. Fong, R. Segawa, S. Powell and J. Ross. 1996a. Methyl Bromide Concentrations in Air Downwind During Aeration of Fumigated Single-Family Houses. HS-1713. March 20, 1996. Worker Health and Safety Branch. Department of Pesticide Regulation. California Environmental Protection Agency, Sacramento, CA.

Gibbons, D.B., H.R. Fong, R. Segawa, S. Powell, and J. Ross. 1996b. Methyl Bromide Concentrations in Air Near Fumigated Single-Family Houses. HS-1717. March 20, 1996. Worker Health and Safety Branch. Department of Pesticide Regulation. California Environmental Protection Agency. Sacramento, CA.

Gotoh, K., T. Nishizawa, T. Yamaguchi, H. Kanou, T. Kasai, M. Ohsawa, H. Ohbayashi, S. Aiso, N. Ikawa, S. Yamamoto, T. Noguchi, K. Nagano, M. Enomoto, K. Nozaki, and H. Sakabe. 1994. Two-Year Toxicological and Carcinogenesis Studies of Methyl Bromide in F344 rats and BDF1 Mice. Proceedings of the Second Asia-Pacific Symposium on Environmental and Occupational Health held 22-24 July, 1993 in Kobe, Japan.

Hallier, E., S. Deutschmann, C. Reichel, H.M. Bolt and H. Peter. 1990. A comparative investigation of the metabolism of methyl bromide and methyl iodide in human erythrocytes. Int. Arch. Occup. Environ. Health. 62(3): 221-225.

Hallier, E., T. Langhof, D. Dannappel, M. Leutbecher, K. Schroder, H.W. Goergens, A. Muller and H.M. Bolt. 1993. Polymorphism of glutathione conjugation of methyl bromide, ethylene oxide and dichloromethane in human blood: influence on the induction of sister chromatid exchanges (SCE) in lymphocytes. Arch. Toxicol. 67(3):173-178.

Hardisty, J.F. 1992. Histopathological Evaluation of Brains From Rats-Inhalation Study of Methyl-Bromide. EPL project number 303-007. DPR Vol. 123-142 #113606.

Haskell, D. 1988a. Methyl bromide fumigation with various commodities. A memorandum dated May 11 to Thomas Thongsinthusak. WH&S, DPR. HSM-98003.

Haskell, D. 1988b. Response to DAR request for additional information regarding frequency and duration of methyl bromide fumigations. A memorandum date August 24 to Thomas Thongsinthusak. WH&S, DAR. HSM-98006.

Hattis, D. and E. Anderson. 1999. What should be the implications of uncertainty, variability, and inherent 'biases'/'conservatism' for risk management decision making? Risk Anal. 19(1):95-107.

Hattis, D. and K. Barlow. 1996. Human interindividual variability in cancer risks--technical and management challenges. Human Ecol. Risk Asssess 2(1):194-220.

Hattis, D. and D.E. Burmaster. 1994. Assessment of variability and uncertainty distributions for practical risk analyses. Risk Anal. 14(5):713-730.

Honma, T., M. Miyagawa, and M. Sato. 1987. Methyl bromide alters catecholamine and metabolite concentrations in rat brain. Neurotoxicol. Teratol. 9(5):369-375.

Honma, T., M. Miyagawa, and M. Sato. 1991. Inhibition of tyrosine hydroxylase activity by methyl bromide exposure. Neurotoxicol. Teratol. 13(1):1-4.

Honma, T., A Sudo, M. Miyagawa, M. Sato, and H. Hasegawa. 1982. Significant changes in monoamines in rat brain induced by exposure to methyl bromide. Neurobehav. Toxicol. Teratol. 4(5):521-524.

Hurtt, M.E., D.A. Thomas, P.K. Working, T.M. Monticello, and K.T. Morgan. 1988. Degeneration and regeneration of the olfactory epithelium following inhalation exposure to methyl bromide: Pathology, cell kinetics, and olfactory function. Toxicol. Appl. Pharmacol. 94(2):311-328.

Jaskot, R.H., E.C. Grose, B.M. Most, M.G. Menache, T.B. Williams, and J.J. Roycroft. 1988. The distribution and toxicological effects of inhaled methyl bromide in the rat. J. Am. Coll. Toxicol. 7(5):631-642.

Kaneda, M., N. Hatakenaka S. Teramoto, and K. Maita. 1993. A two-generation reproduction study in rats with methyl bromide-fumigated diets. Food Chem. Toxicol. 31(8):533-542.

Kaneda, M., H. Hojo, S. Teramoto, and K. Maita. 1998. Oral teratogenicity studies of methyl bromide in rats and rabbits. Food Chem. Toxicol. 36(5):421-427.

Khera, K.S. 1984. Maternal toxicity: a possible factor in fetal malformations in mice. Teratology 29(3):411-16.

Khera, K.S., H.C. Grice HC, and D.J. Clegg, eds. 1989. Current Issues in Toxicology, Interpretation and Extrapolation of Reproductive Data to Establish Human Safety Standards. New York: Springer.

Kornburst, K.S., and J.S. Bus. 1983. The role of glutathione and cytochrome P-450 in the metabolism of methyl chloride. Toxicol. Appl. Pharmacol. 67(2):246-56.

Kramers, P.G., C.E. Voogd, A.G. Knaap, and C.A. van der Heijden. 1985. Mutagenicity of methyl bromide in a series of short-term test. Mutat. Res. 155(1-2):41-47.

McGregor, D.B. 1981. Tie II Mutagenic Screening of 13 NIOSH Priority Compounds. National Institute for Occupational Safety Health. Inveresk Research Int. Ltd. DAR Vol. 123-103 #66718, #66719, #66720, #66721, and #66722.

Medinsky, M.A., J.A. Bond, J.S. Dutcher, and L.S. Birnbaum. 1984. Disposition of [^{14}C] methyl bromide in Fischer-344 rats after oral or intraperitoneal administration. Toxicology. 32(3):187-196.

Medinsky, M.A., J.S. Dutcher, J.A. Bond, R.F. Henderson, J. L. Mauderly, M.B. Snipes, J.A. Mewhinney, Y.S. Cheng, and L.S. Birnbaum. 1985. Uptake and excretion of [14C] methyl bromide as influenced by exposure concentration. Toxicol. Appl. Pharmacol. 78(2):215-225.

Mertens, J.J.W.M. 1997. A 24-Month Chronic Dietary Study of Methyl Bromide in Rats. Laboratory Study no. WIL-49014. WIL Research Laboratories. DAR Vol. 123-179 #158746.

Michalodimitrakis, M.N., A.M. Tsatsakis, M.G. Christakis-Hampsas, N. Trikillis, and P. Christodoulou. 1997. Death following intentional methyl bromide poisoning: toxicological data and literature review. Vet. Hum. Toxicol. 39(1):30-34.

Mitsumori, K., K. Maita, T. Kosaka, T. Miyaoka, and Y. Shirasu. 1990. Two-year oral chronic toxicity and carcinogenicity study in rats of diets fumigated with methyl bromide. Food Chem. Toxicol. 28(2):109-119.

Moriya, M., T. Ohta, K. Watanabe, T. Miyazawa, K. Kato, Y. Shirasu. 1983. Further mutagenicity studies on pesticides in bacterial reversion assay systems. Mutat. Res. 116(3-4):185-216.

Mortelmans, K.E., and G.F. Shepherd. 1980. In Vitro Microbiological Mitotic Recombination Assay of Methyl Bromide Using S. cereisiae D3. SRI International. DAR Vol.123-044 #913095.

NRC (National Research Council). 1991. Human Exposure Assessment for Airborne Pollutants: Advances and Opportunities. Washington, DC: National Academy Press.

NRC (National Research Council). 1993. Pesticides in the Diets of Infants and Children. Washington, DC: National Academy Press.

NRC (National Research Council). 1994. Science and Judgement in Risk Assessment. Washington, DC: National Academy Press.

Nelson, H.H., J.K. Wiencke, D.C. Christiani, T.J. Cheng, Z.F. Zuo, B.S. Schwartz, B.K. Lee, M.R. Spitz, M. Wang, X. Xu, et al. 1995. Ethnic differences in the prevalence of the homozygous deleted genotype of glutathione S-transferase theta. Carcinogenesis 16(5):1243-1245.

Newton, P.E. 1994a. An Up-and-Down Acute Inhalation Toxicity Study of Methyl Bromide in the Dog. Study no. 93-6067 prepared by Pharmaco LSR, East Millstone, NJ for Chemical Manufacturers Association, Washington, DC.

Newton, P.E. 1994b. A Four Week Inhalation Toxicology Study of Methyl Bromide in the Dog. Study no. 93-6068 prepared by Pharmaco LSR, East Millstone, NJ for Chemical Manufacturers Association, Washington, DC.

Newton, P.E. 1996. A Chronic (12-Month) Toxicity Study of Methyl Bromide Fumigated Feed in the Dog. Study no. 94-3186. Huntingdon Life Science. DAR Vol. 123-175. #143945.

NTP. (National Toxicology Program). 1992. Toxicology and Carcinogenesis Studies of Methyl Bromide (CAS No. 74-83-9) in B6C3F1 Mice (Inhalation Studies). Tech. Rep. Series No. 385. Research Triangle Park: U.S. Department of Health and Human Services, Public Health Service, National Institutes of Health.

OEHHA (Office of Environmental Health Hazard Assessment). 1993. Proposition 65: Safe Use Determination Workshop for Methyl Bromide (Tuesday November 30, 1993.) Transcript. California Environmental Protection Agency, Office of Environmental Health Hazard Assessment. 205 pp.

OEHHA (Office of Environmental Health Hazard Assessment). 1999. Comments on the Department of Pesticide Regulation's Draft Risk Characterization Document for Inhalation Exposure to the Active Ingredient Methyl Bromide. From Anna M. Fan, California Office of Environmental Health Hazard Assessment, to Gary T. Patterson, California Department of Pesticide Regulation, September 1, 1999 (Appendix L of DPR 1999).

Overstreet, D.H., R.W. Russell, B.J. Vasquez, and F.W. Dalglish. 1974. Involvement of muscarinic and nictonic receptors in behavioral tolerance to DFP. Pharmacol. Biochem. Behav. 2(1):45-54.

Peter, H., S. Deutschmann, C. Reichel, and E. Hallier. 1989. Metabolism of methyl chloride by human erythrocytes. Arch. Toxicol. 63(5):351-5.

Putnam, D.L. and M.J. Morris. 1991. Micronucleus Cytogenetic Assay in Mice. Microbiological Assoc. Inc. DAR Vol. 123-136 #99090.

Raabe, O.G. 1986. Inhalation Uptake of Selected Chemical Vapors at Trace Levels. University of California, Davis, CA. Submitted to The Biological Effects of Research Section, California Air Resources Board, Sacramento, CA.

Raabe, O.G. 1988. Retention and Metabolism of Toxics. Inhalation uptake of Xenobiotic Vapors by People. University of California, Davis, CA. Submitted to The Biological Effects Research Section, California Air Resources Board, Sacramento, CA.

Reuzel, P.G.J., C.F. Kuper, H.C. Dreef-van der Meulen, V.M.H. Hollanders. 1987. Chronic (29-Month) Inhalation Toxicity and Carcinogenicity Study of Methyl Bromide in Rats. Report No. V86.469/221044. Zeist, The Netherlands: TNO-CIVO Toxicology and Nutrition Institute.

Reuzel, P.G.J., H.C. Dreef-van der Meulen, V.M.H. Hollanders, C.F. Kuper, V.J. Feron, and C.A. and van der Heijden. 1991. Chronic Inhalation Toxicity and Carcinogenicity Study of Methyl Bromide in Wistar Rats. Food Chem. Toxicol. 29(1):31-39.

Rosenblum, I., A.A. Stein, and G. Eisinger. 1960. Chronic ingestion by dogs of methyl bromide-fumigated food. Arch. Environ. Health 1:316-323.

Segawa, R., T. Barry, and B. Johnson. 2000a. Recommendations for Methyl Bromide Buffer Zones for Field Fumigations. Memo to John S. Sanders, Chief, Environmental Monitoring and Pest Management Branch, Department of Pesticide Regulation, January 21, 2000.

Segawa, R., B. Johnson, and T. Barry. 2000b. Summary of off-site air monitoring for methyl bromide field fumigations. Memo to John S. Sanders, Chief, Environmental Monitoring and Pest Management Branch, Department of Pesticide Regulation, January 21, 2000.

Seiber, J.N. 1999. Letter to Douglas Y. Okumura, Acting Assistant Director, Division of Enforcement, Environmental Monitoring and Data Management, Department of Pesticide Regulation. Comments on report, Evaluation of Charcoal Tube and SUMMA Canister Recoveries for Methyl Bromide Air Monitoring, from James N. Seiber, University of Nevada, May 5, 1999.

Sikov, M.R., W.C. Cannon, D.B. Carr, R.A. Miller, L.F. Montgomery, and D.W. Phelps. 1981. Teratologic Assessment of Butylene Oxide, Styrene Oxide and Methyl Bromide. Contract no. 210-78-0025. Battelle, Pacific Northwest Lab. Submitted to the Division of Biomedical and Behavioral Science, National Institute for Occupational Safety and Health, U.S. Department of Health and Human Services, Cincinnati, OH.

Simmon, V.F., K. Kauhanen, R.G. Tardiff. 1977. Mutagenic activity of chemicals identified in drinking water. Pp. 249-258 in: Progress in Genetic Toxicology, D. Scott, B.A. Bridges, and F.H. Sobels, eds. Amsterdam: Elsevier/North Holland Biodmedical Press. DAR Vol 123-109 #87801.

Stadler, J., M.J. Kessedjian, and J. Perraud. 1983. Use of the New Zealand white rabbit in teratology: Incidence of spontaneous and drug-induced malformations. Food Chem. Toxicol. 21(5):631-636.

Thompson, K.M. 1999. Developing univariate distributions from data for risk analysis. Human Ecol. Risk Assess. 5(4):755-783.

Thongsinthusak, T., D. Haskell, and J. Ross. 1999. Estimation of exposure of persons to methyl bromide during and/or after agricultural and non-agricultural uses. California Environmental Protection Agency, Department of Pesticide Regulation. HS-1659 (Appendix F of DPR 1999).

Tyl, R. 1991. Comments on Methyl Bromide Inhalation Teratology Study in New Zealand White Rabbits, W.J. Breslin, C.L. Zablotny, G.J. Bradley, and L.G. Lomax, Toxicology Research Laboratory, Health and Environmental Sciences, The Dow Chemical Company, Midland, MI: Study No. K-000681-033, Final Report, June 18, 1990 by R.W. Tyl, Senior Program Director, Research Triangle Institute to Dr. R. Franklin Handy, Methyl Bromide Industry Panel, Great Lakes Chemical Corporation, dated February 25, 1991. 31 pp.

Appendix A

Biographical Information on the Subcommittee for the Review of the Risk Assessment of Methyl Bromide

CHARLES H. HOBBS (Chair) is director of the Toxicology Division at the Lovelace Respiratory Research Institute. He received his D.V.M. from Colorado State University. His research focuses on the long-term biological effects of inhaled materials and the mechanisms by which they act. He is a diplomate of the American Board of Toxicology and certified in general toxicology. Dr. Hobbs serves as a member of the Committee on Toxicology and previously served on the Committee on Toxicological and Performance Aspects of Oxygenated fuels.

JANICE E. CHAMBERS is professor and director of the Center for Environmental Health Sciences in the College of Veterinary Medicine at Mississippi State University. She received her Ph.D. in animal physiology from Mississippi State University. Her research focuses on neurotoxicology of insecticides including neurochemical and behavior studies and insecticide metabolism. She is a diplomate of the American Board of Toxicology. Dr. Chambers previously served as a member of the NRC's Panel on Life Sciences for postdoctoral fellowships.

FRANK N. DOST is professor emeritus from the Department of Agricultural Chemistry at Oregon State University and affiliate professor in the Department of Environmental Health at the University of Washington. He received his D.V.M. from Washington State University. Dr. Dost's research interests include the estimation of environmental and occupational chemical exposure and risk assessment and the metabolic fate of toxicants. Previously, Dr. Dost served on the NRC committee on toxicology of hydrazines.

DALE B. HATTIS is research professor in the Center for Technology, Environment, and Development at Clark University. He received his Ph.D. in genetics from Stanford University. His research focuses on the development and application of methodologies to assess the health impacts of regulatory options with an emphasis on incorporating interindividual variability data into risk assessments for both cancer and non-cancer endpoints. Previously, Dr. Hattis was a member of the NAS/IOM Committee on Evaluation of the Safety of Fishery Products and the NRC Committee on Neurotoxicology and Risk Assessment.

MATTHEW C. KEIFER is co-director of the Pacific Northwest Agricultural Safety and Health Center and director of the occupational medicine program at the University of Washington. He received his M.D. from the University of Illinois and his M.P.H. from the University of Washington. Dr. Keifer's research interests focus on the health of agricultural workers with specific focus on the health effects of occupational pesticide exposure. He is a diplomate of the American Board of Internal Medicine.

ULRIKE LUDERER is assistant professor with the Center for Occupational and Environmental Health at the University of California at Irvine. She received her M.D. and Ph.D. from Northwestern University and her M.P.H. from the University of Washington. Dr. Luderer's research focuses on reproductive effects and neuroendocrine alterations as a result of exposure to environmental toxicants, particularly volatile organics. She is a diplomate of the American Board of Internal Medicine.

GLENN C. MILLER is director of the Center for Environmental Sciences and Engineering at the University of Nevada, Reno. He received his Ph.D. in Agricultural Chemistry from the University of California at Davis. Dr. Miller's research focuses on the fate and transport of airborne pesticides following major uses and the effects of deposited residues on soils including their photodegradation.

SYLVIA S. TALMAGE is a toxicologist in the Life Sciences Division at Oak Ridge National Laboratory. She received her Ph.D. in ecology/environmental toxicology from the University of Tennessee. Dr. Talmage's research focuses on the sources, fate, and toxicity of chemical warfare agents. She is a diplomate of the American Board of Toxicology and certified in general toxicology.

Appendix B

Public Access Materials

The following materials (written documents) were made available to the committee at or after its first meeting, October 4, 1999, Beckman Center:

1. California Environmental Protection Agency, Department of Pesticide Regulation. 1999. Methyl Bromide: Risk Characterization Document for Inhalation Exposure. Draft. March 1, 1999. 149 pp. with 9 appendices.

2. Memorandum from Lori Lim and Stephen Rinkus, California Department of Pesticide Regulation to Gary Patterson, California Department of Pesticide Regulation. Subject: Methyl Bromide Assignment #98-0507. Dated September 25, 1998. 23 pp.

3. Methyl Bromide Industry Panel. 1998. Toxicological Endpoint Evaluation and Exposure Assessment for Methyl Bromide. August 18, 1998. 33 pp. With 2-pg cover letter from David Weinberg to James Wells, Department of Pesticide Regulation.

4. Comments on the Department of Pesticide Regulation's Draft Risk Characterization Document for Inhalation Exposure to the Active Ingredient Methyl Bromide. From Anna M. Fan, California Office of Environmental Health Hazard Assessment, to Gary T. Patterson, California Department of Pesticide Regulation, dated September 1, 1999. 21 pp.

5. Risk Assessment of Methyl Bromide. Presented by Lori O. Lim and Thomas Thongsinthusak, California Department of Pesticide Regulation. October 4, 1999. 29 pp.

6. Chemical Manufacturers Association, Methyl Bromide Industry Panel. Presented by Vincent Piccirillo, NPC, Inc. October 4, 1999. 28 pp.

7. Methyl Bromide Use in California: Public Health Concerns for Residents near Fumigated Agricultural Fields. Presented by Bill Walker, Environmental Working Group. October 4, 1999. 135 pp.

8. Public Health Concerns in the Methyl Bromide Reassessment. Presented by Amy Kyle, Consulting Scientist for the California Rural Legal Assistance Foundation. October 4, 1999. 19 pp.

9. California Environmental Protection Agency, Department of Pesticide Regulation. 1999. Methyl Bromide: Risk Characterization Document for Inhalation Exposure. Draft. October 1999. 467 pp.

10. Letter from James N. Seiber, University of Nevada, to Douglas Okumura, California Department of Pesticide Regulation, dated May 5, 1999. Comments on report "Evaluation of Charcoal Tube and SUMMA Canister Recoveries for Methyl Bromide Air Monitoring." 54 pp.

11. Heinz's responses to Jim Seiber's comments. Draft. Undated. 3 pp.

12. Mini-Memo from Terri Barry to Kean Goh, dated May 19, 1999. Draft. Responses to comments on statistical aspects of the report entitled "Evaluation of Charcoal Tube and SUMMA Canister Recoveries for Methyl Bromide Air Sampling." 4 pp.

13. U.S. EPA. 1991. Guidelines for Developmental Toxicity Risk Assessment. U.S. Environmental Protection Agency, Office of Research and Development, Risk Assessment Forum. EPA/600/FR-91/001. 67 pp.

14. Letter from Courtney Price, Vice President CHEMSTAR, on behalf of the Chemical Manufacturers Association's Methyl Bromide Industry Panel to Dr. Charles Hobbs, NRC Subcommittee on Methyl Bromide. Dated November 8, 1999. 8 pp. with 2 attachments of published article by Medinsky et al. (1985) and bar chart.

PUBLIC ACCESS MATERIALS   89

15. California Rural Legal Assistance Foundation. 1999. Letter From Anne Katten and J. Felix de la Torre of the CRLAF to Roberta Wedge, NRC, regarding review of California Department of Pesticide Regulation Risk Characterization for Methyl Bromide. Dated December 23, 1999. 2 pp. With attachment "Technical Comments of California Rural Legal Assistance Foundation: The NAS Review of the CDPR Methyl Bromide Risk Characterization, December 1999." 10 pp.

16. Letter from Gary T. Patter, Division of Registration and Health Evaluation, California Department of Pesticide Registration to Roberta Wedge, NRC, regarding an issue paper submitted by the Methyl Bromide Industry Panel (MBIP) on the pharmacokinetics and metabolism of methyl bromide. Dated January 7, 2000. 2 pp. With 3 attachments including; 1) the issue paper, 2) review and comments of the issue paper by the DPR staff; and 3) questions posed by DPR to the MBIP at a meeting where the issue paper was presented.

17. Letter from Jodi Kuhn, Methyl Bromide Industry Panel (MBIP) to Roberta Wedge, NRC with comments from the MBIP to the California Department of Pesticide Regulation. Dated January 21, 2000. 1 pg. With an attached letter from MBIP to Paul Helliker, CDPR, dated January 11, 2000 (1 pg) and a 3 page attachment entitled "Methyl Bromide: Supplemental Information on Metabolism."

18. Memorandum from Lori Lim, Department of Pesticide Regulation (DPR), to Gary Patterson, DPR, regarding methyl bromide assignment #98-0408. Dated July 3, 1998. 2 pp. With an attached memorandum from Linnea J. Hansen, U.S. Environmental Protection Agency, Office of Prevention, Pesticides, and Toxic Substances, Health Effects Division, to Margaret Stasikowski, Health Effects Division, entitled "Methyl Bromide: Review of Draft Toxicology and Hazard Identification Document Prepared by the Department of Pesticide Regulation, California Environmental Protection Agency. Dated June 11, 1998. 6 pp.

19. Letter from Cindy Tuck, Law Offices of William Thomas, Sacramento, CA, to Roberta Wedge, NRC, regarding methyl bromide recovery rate: new document for review by NRC subcommittee on methyl bromide. With an attached memorandum from Jay Gan, ARS USDA, to Dr. Duafain on DPR study and methyl bromide recovery rates. Dated March 21, 2000. 5 pp.

Appendix C

Calculation of Air Exchange Rates

Air exchange rates are defined in terms of a general one-compartment model of air exchange with immediate and perfect mixing of air inside residences. For any contaminant in the assumed well-mixed pool of air in the living spaces, this leads to an expectation of simple exponential decline of air concentrations with time:

$$C(t) = C(0)e^{-kt},$$

where $C(0)$ is the initial concentration of the contaminant inside the house, $C(t)$ is the concentration of the contaminant at any specific time after $t = 0$, and k is a rate constant in units of reciprocal time (i.e., if time is expressed in hours, k is in reciprocal hours, or, by convention, "air changes per hour"). The relationship between the rate constant k and the half-life (the time required to reduce the air concentration by half) is easily derived by setting $C(t)$ to one-half of $C(0)$:

$$C(t_{1/2}) = .5C(0) = C(0)e^{-kt_{1/2}},$$

After the cancellation of the $C(0)$'s, and taking the natural logarithm of both sides of the equation:

$$\ln(.5) = -kt_{1/2}$$

$$t_{1/2} = \ln(2)/k \quad \text{or} \quad k = \ln(2)/t_{1/2}$$

ILLUSTRATIVE LOGNORMAL TREATMENT OF DATA FOR SELECTED OCCUPATIONAL EXPOSURES

Figure A-1 shows lognormal probability plots of the individual data points for several groups of workers in the shallow-shank tarp method application of methyl bromide. In this type of plot, correspondence of the points to the re-

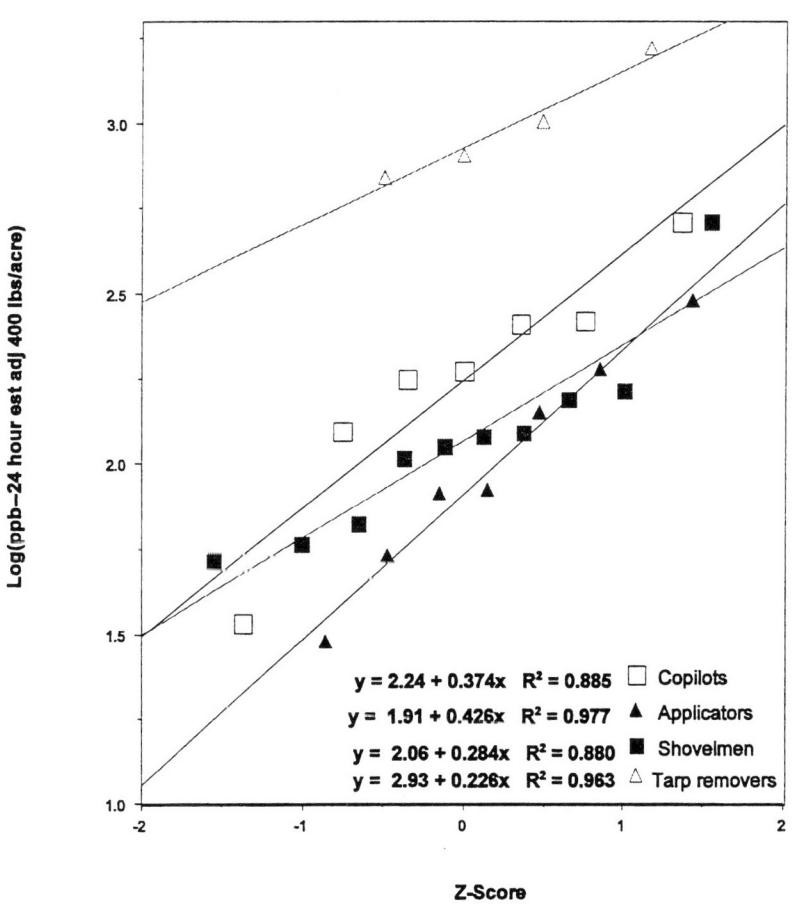

FIGURE A-1

gression line is a quick qualitative indicator of the degree to which the data points are well described by the chosen distribution. In these cases, the fits are far from perfect, suggesting some possible heterogeneity in the data, but the lognormal plots in Figure A-1 are generally better than corresponding normal distribution fits (Figure A-2). For these same worker groups, Table A-1 below compares the reported highest observed values with 95th percentile values calculated from the fitted normal and lognormal distributions. In general, the lognormal fits project somewhat higher 95th percentiles than the normal fits.

TABLE A-1 Comparison of Observed Values with 95th Percentile Values

Occupational Group	Number of Data Points (Including Non-Detects)	Highest Observed Acute (24 hr) Exposure	95th Percentile Calculated from Normal Fit	95th Percentile Calculated from Lognormal Fit
Copilots	7	518	479	716
Applicators	8	303	293	408
Shovelmen	10	515	330	337
Tarp Removers	5	1659	1820	1990

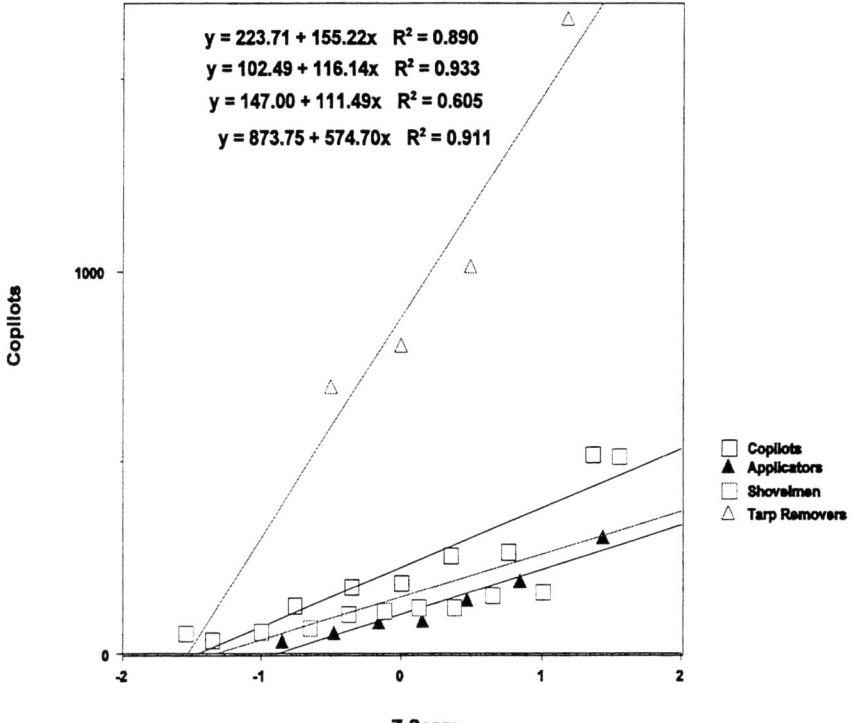

Figure A-2